FELIPE SABOYA DE SANTA CRUZ

MKTECH

VOLUME 1

INVASION AND MIND CONTROL

FELIPE SABOYA DE SANTA CRUZ ABREU

MKTECH

VOLUME 1

INVASION AND MIND CONTROL

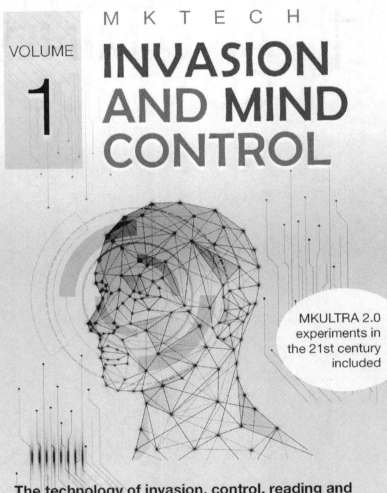

MKULTRA 2.0 experiments in the 21st century included

The technology of invasion, control, reading and torture of the mind that will change humanity forever.

2021 © Felipe Saboya de Santa Cruz Abreu

TITLE: MKTECH: Invasion and Mind Control - Volume 1
1st edition – October 2019

Author: **Felipe Saboya de Santa Cruz Abreu**
felipessca@gmail.com
Diagrams: **Felipe Saboya de Santa Cruz Abreu**
Text formatting: **Felipe Saboya de Santa Cruz Abreu**

Illustrations: **Eloy Rondon**
eloyartes@hotmail.com

Translation: **Luene Langhammer Alves**
luene.langhammer.alves@gmail.com

Book design: **Rubens Lima**
https://capista.com.br/
Book cover illustration: Freepik.com

ISBN: 9798798603299

All rights reserved and protected under Brazilian Law number 9610 from 02/19/1998.
No part of this publication may be reproduced or transmitted in any form or by any means, electronic or mechanical, including photocopying, recording or any information storage and retrieval system, without permission in writing from the author.
The rights of this book belong to the author/editor FELIPE SABOYA DE SANTA CRUZ ABREU.

Official website:

www.invasionandmindcontrol.com

Acknowledgments

I have to start by thanking my family who gave me unconditional support to research and write this book. Thank you to everyone who helped me without even knowing it. To the people who passed through my life and contributed, in some way, to the completion of this book. The everyday coincidences of life, situations that seemed to be the most adverse in the end became the solution for the best insights I had.

To all my teachers and professors who indirectly participated in this journey by helping me to keep my body and mind healthy with daily training. To the events that occurred as if they had been planned and predicted, thus enabling the choices to converge at this unique moment.

To longtime friends.

Thank you all.
Felipe Saboya de Santa Cruz Abreu

Introduction — 11

CHAPTER 1 — 15

WHAT IS THOUGHT AND HOW DO WE THINK? — 15
1.1 - What is Language? — 17

CHAPTER 2 MKTECH — 25
the electronic systems that FORM the Mind Control Technology — 25

CHAPTER 2.1 — 27

EMR - ELECTRONIC MIND READING — 27
Meet EMR — 28
Hacking of profiles and passwords from computers, websites and systems — 31
Intellectual property theft — 33
Theft of Confidential Information — 35
Theft of credit cards and PINs — 36
ATM withdrawals without biometric fingerprint technology — 36
End of interpersonal relationships — 37
End of privacy — 38
End of the right to free thinking — 38
2.1.1 - How does the whole process of amplifying brain waves and extracting their contents from a remote position take place? — 40
2.1.2 - Brain Functioning and Electromagnetism — 46
Speaking, reading and writing — 49
Summing up — 54
2.1.3 - The vinyl record — 58
Theft of vocalized thought and visual thinking — 59

CHAPTER 2.2 — 63

V2K – "VOICE TO SKULL", INTRACRANIAL VOICE, MICROWAVE VOICE OR MICROWAVE HEARING EFFECT — 63
But what is a radar? — 64
2.2.1 - The history of the Intracranial Voice — 65
2.2.2 - How the microwave voice affects hearing and the brain — 68
Program interface — 79
2.2.3 - How is the microwave voice demodulated by the brain? — 84

CHAPTER 2.3 — 87

SYNTELE - SYNTHETIC ELECTRONIC TELEPATHY — 87

SYNTELE for torture, experiments and information theft … 95
I – Battle tips for the Targeted Individual … 107
2.3.1 - SYNTELE and Electronic Schizophrenia … 109
Symptoms … 110

CHAPTER 2.4 — 119

EMRa – ELECTRONIC MIND READING (AUDITORY) — 119
But how does it happen? … 127
2.4.1 - Thalamus … 127
Vocalized thoughts … 132
V2K microwave voice capture … 134
2.4.2 - How the algorithm that filters conversations works … 135

CHAPTER 2.5 — 137

D2K – Synthetic Electronic Dream — 137
2.5.1 - Gamification of dreams … 146
2.5.2 - What is sleep and dream? How does D2K affect each stage? … 153
Dreams … 153
Function of dreams in REM sleep … 162
2.5.3 - Dreams created by D2K in REM sleep … 163
Waking up the target by synchronizing with dreams … 173
Gap within dreams … 182
2.5.4 - How do memory, remembrance and imagination work? … 185
Memories … 185
Remembrances … 187
II - Important tips for the Targeted Individual … 188
2.5.5 - Creating long-term memories using D2K (BYPASS) by deceiving the brain. … 189
Mixing reality with dreams and dreams with reality. *Déjà vu and Déjà Rêvé* … 190
Creating the "Manchurian Candidate" … 193
"Bypass" or alternate path to insert commercial images … 193
III - Important tips for the Targeted Individual … 196
IV - Important tips for the Targeted Individual … 200
V – Important tips for the target … 201
Exploring the sleep transition state … 201
2.5.6 - Dream management using V2K … 203
2.5.7 - Sex-related dreams and testing of indiscreet situations … 205
"Dates" or pornographic encounters … 209
Inserting familiar faces in dreams … 211
REM rebound … 214
2.5.8 - Creating test models about the target … 215
2.5.9 - "Tunguska Sound" or Night Bang … 219
2.5.10 - Dislocated Thought or Dissonant Mind … 221
2.5.11 - Altered time perception … 223
2.5.12 - But, after all, how is it possible to replace dreams? … 223

Conclusion 227

CHAPTER 2.6   229
DANGER IN THE USE OF THE TECHNOLOGY (PART 1) - A BOY CALLED JAMES   229
What really happened to the boy?   233

CHAPTER 2.7   235
V2K - sound within another sound   235

CHAPTER 2.8   249
DANGER IN THE USE OF THE TECHNOLOGY (PART 2) - Mayday! Mayday! Mayday! Danger to civil passenger aviation 249

CHAPTER 2.9   256
DANGER IN THE USE OF THE TECHNOLOGY (PART 3) - "WINTER SOLDIER" ESTEBAN SANTIAGO   256
Esteban Santiago, 26   257
"Hallucinations"?!   257
A Brief Reflection   259

CHAPTER 2.10   261
EMRo - ELECTRONIC MIND READING (optical)   261
The sight   262
2.10.2 - Visual memories of thought   267

CHAPTER 3 - PHYSIOLOGICAL MONITORING - NEURAL BIOMETRIC SIGNATURE - REMOTE POLYGRAPH   271

RNM - Remote Neural Monitoring   271
3.1- Neural Remote Biometrics   272
3.2 - Telemetric EEG, Electronic Brain Link, Remote Neural Monitoring   278
3.2.1 - Body position in space   282
3.2.2 - Brain waves   287
Motivations and intentions   287
Factor 1: Primary emotions: 288
Factor 2: States of consciousness: 288
3.2.3 - Target location anywhere in the world. Neural GPS   292
3.3 - Remote Polygraph - The most efficient lie detector ever created   293
How normal polygraph tests work   295
Experiments captured in the face of the remote polygraph   300
3.4 - How can they see me all the time, inside my house, in the bathroom, in the bedroom, at work, at the beach, in the park, on the street, at

friends' houses, in distant places in the countryside, at my cousin's or sister's house? How do they "see" everything I do everywhere I go? They see what I see and they hear what I hear all the time 304
The technology 305

Epilogue 307
Glossary I 309
Glossary II 311
About the Author 312
References 313

Introduction

"Nothing is too wonderful to be true, if it be consistent with the laws of nature."

— Michael Faraday.

High technology, cover-up, disinformation and destruction of human lives. The technology of direct manipulation and remote physical interaction of thoughts, previously unknown to the general public and which was exclusive to developed countries, has already arrived in Brazil and has caused havoc wherever it goes. The Mind Control Technology (MKTECH) is an advanced electromagnetic weapon that interferes with the bioelectric functioning of the brain of any evolved animal on the planet, such as mammals — and that includes us, human beings.

Welcome to the best kept secret of the last 60 years. It's by far the most important one since the Manhattan Project, a period in which the discovery of nuclear fission and the development of the atomic bomb changed civilization. Today the technology of control, invasion, mental and psychological torture has been doing the same thing. So, prepare to embark on the path of profound transformations in society. We're going to deal with new paradigms never considered in history, which includes the end of cognitive privacy and the content of our thinking.

In this book, we're going to discuss how technology and science are advanced to the point that electronic equipment is directly capable of interacting with the functioning of the brain and its bioelectric functions, all of which is conducted remotely using electromagnetic waves. Moreover, we're going to focus on the people or groups of people who use this technological arsenal to torture, defraud, steal and murder all over the world — or just to have fun with the suffering of others.

All the topics covered in this book are the result of careful research based on scientific approach. They're the outcome of seven years of direct contact with this technology in full operation, following human test subjects and their suffering day by day and unraveling a complex, intricate and macabre system that involves disturbing phenomena and extensive reflections to understand them in order to show the danger to which we are exposed today. We're going to cover the reading of thoughts directly from the brain, theft of information and intellectual property to torture, murder and fraud in civil service competitive examinations.

Microwave antennas, satellites with instruments and sensors specialized in hacking the human brain and invading the mind using several advanced systems, programs specialized in analyzing and monitoring brain waves and their content, thus creating from a distance the most potentially destructive weapon ever developed by humans.

The book is also intended to alert readers to the arrival of this technology and its consequences on society as a whole, as well as the direct implications for humanity's basic social interactions. It's a complete violation of our constitutional rights perpetrated by these electronic devices. In their essence, they're modern weapons of the 21st century capable of reaching thousands of people at the same time without them being aware of what is happening or even causing them to draw incorrect conclusions about such an event when perceived.

As this technology has been in the dark since it was conceived, with restricted access to the military environment and intelligence institutions like CIA, KGB, MI6, BND, MOSSAD, MSS, among others, it's extremely complicated for people to believe that this type of event is real. Even people with advanced knowledge in science are unable to form a general picture of what is going on in the world lately.

Based on that, I'm going to prepare the reader for what is happening behind the scenes of modern society, since we're facing a reality that is difficult to adapt; a fact that will change practically everything that we establish as societies and their foundations: that it is indeed possible to remotely monitor the electrical brain activity of entire regions, or selected parts, through the transmission and reception of electromagnetic waves at several different intervals, which interact directly with the human mind of any individual on the planet. Thus, the ability to read the content of thoughts, to hear vocalized thoughts, to see visual thoughts from visual memories — which opens unique precedents — becomes possible.

Furthermore, they're massacring people across the Earth using the voice-to-skull technology (V2K), a terrible device that inserts sounds and voices directly into an individual's brain, driving them insane. As if that were not enough, there is still a powerful weapon capable of completely replacing a person's dreams while sleeping, as if they were a TV broadcast or a computer game, causing unprecedented disruption. This is happening right now with thousands of people.

We're going to discuss these technologies that together make up the Mind Control Technology (MKTECH). We're going to differentiate technologies in terms of their functionalities that serve for surveillance, espionage and unauthorized access to thoughts, and how they're being used for torture and murder. Then we're going to understand the role of the organized crime that was born along with this technology, as well as the techniques used to accomplish shady, hidden goals. Finally, we're going to get to know MKULTRA 2.0, which provides the infrastructure and protocol to carry out attacks on victims, as its namesake — called MK-ULTRA— did in 1950.

This book shows the reality behind this technology hidden from most people, but which has become popular in Brazil and worldwide. We're going to know the most nefarious weapon ever conceived by humans after the nuclear bomb. It's invisible, can modify and interfere physically and remotely in the mental content of living beings, altering the entire cognitive process, modifying dreams when we're in deep sleep, changing behaviors, reaching thousands of people without leaving a trace and completely transforming society. This is the transformation of human lives at the push of a button!

CHAPTER 1
WHAT IS THOUGHT AND HOW DO WE THINK?

"I think, therefore I am."

— René Descartes.

Since the beginning of mankind, humans have attempted to understand how the world around them works, in what context we fit into it and how we establish the connection between reality and the brain. The brain, in turn, interprets all this information and synthesizes it in a model of the external world so that we can understand it internally, and which generates profound questions about the existence and the way we coordinate and organize this information.

From the 19th century onwards, the separation of thought and consciousness — the continuous, dynamic flow of ideas, judgment, concepts, among others — was created. Even though there was still no comprehensive theory in the 20th century that managed to conceptualize the reason for thinking the way we think, some divisions were established by several authors, such as: subjectivity, which refers to personal experiences; judgment, which is the process of establishing relationships between concepts; and the content, which is thinking about something, the content itself.

We generally don't think about the particulars of the cognitive processes that create thoughts or how they work. On a daily basis, we only use the resources offered by the brain to communicate, to reason and to establish relationships between abstract and subjective concepts that use memory and deal with external stimuli from the environment. But, after all, what is thought?

Thought is a brain activity that is structured in language and correlates the individual with themselves, with others and with the environment. It's an intellectual activity that brings into existence the meaning, the understanding of information and the organization of external stimuli that we receive at all times, be it visual, auditory or tactile stimulus.

It can be argued that it's a product of the mind that arises through rational activities of the intellect or by abstractions of the imagination, thus helping us to form consciousness. It's the capacity to conceive, combine and compare ideas. To think is to dialogue, to converse, to ramble, to use the senses in an internal way in a series of rational operations, such as analysis, synthesis, comparison, generalization and abstraction. We think by seeing internal images, hearing sounds or speaking internally. The process involves a series of neural networks that together build the thought that later can be communicated through speech, writing or sign language.

Today, the concept of thought structure has become broader and at the same time more specific than in the past. It's organized into logical, rational thinking, which consists of the flow of ideas, symbols and associations directed to an object through attention. It encompasses the process of judgment, understanding, reasoning and anticipation of facts. The main intellectual components of thought are then divided into: concept, judgment and reasoning.

Concept is a purely verbal scheme that encompasses in a single mental operation the relationship between species and genus. It consists of the abstractions of memories derived from the repetition of a constant stimulus associated with the denial of unnecessary stimuli, separating the fundamental and the circumstantial through generalization and abstraction of objects.

Judgment is the most complex product of the intellect. It makes use of logic to establish associations between different concepts; it's the result of the individual's judgment of objective reality. Through internalized beliefs and associative processes, we evaluate the sensory data in order to position us in the world.

Logical thinking leads to judgment, and the relationship between judgments constitutes reasoning. It oscillates between abstract and fantasy thinking, with no determined direction. There is, in fact, no concept considered the most accurate for the definition of thought. It can be said that thinking is the recognition of the coherent flow of ideas that occur naturally and effortlessly. Thus thought, which is the product of the mind, produce a range of abstract concepts that designate the mind itself.

Within this comprehensive philosophical concept that is thought we can recognize two distinct types that correlate in order to create meanings, which are part of the composition of thought. They're called vocalized thought and visual thinking. There can be thought only with images, however, there is no vocalized thought without language.

The complexity of thought embraces several areas of the brain. So, in order to delimit its scope, we're going to focus on two different types of thinking: those based on images (visual memory) and vocalized thoughts, which essentially depend on the language learned and the word. To understand how the Mind Control Technology (MKTECH) works, we have to pay attention to the cortical dynamics that generate these thoughts. In order to mainly understand how vocalized thought and visual thinking work, we cannot dissociate or study language separately and its influencing power in the rest of the cortex, as well as its importance in these specific processes. Visual thinking is also associated with language and words. However, vocalized thoughts depend essentially on the language.

So, let's start thinking about these concepts to understand how this terrible technology works.

1.1 - What is Language?

Language is an intellectual competence that doesn't depend on physical objects in the world, but on the exploration of auditory channels. It's a very complex cognitive/psychological function, involving several elements, such as social communication and intellectual activity. Language is the most used means of communication in society, capable of transferring information from the interlocutor to the listener. It may or may not be expressed through speech, that is individual and in turn is linked to the language (idiom). Once learned, it spends most of the time in silent activity. Language gives the individual the ability to talk about things and facts located remotely in space and time. It allows them to solve problems outside the momentary physical situation.

Language conveys the concepts, judgments and reasoning of thought. It's able to transmit emotions, or a set of them, and activate those same emotions in the receivers through a grouping of sounds in sequence, rhythmically organized and metrically constructed. In the Western world, reading depends on areas of the brain that process linguistic sounds; in the Eastern system, ideographic reading depends crucially on the centers of pictorial materials. Japanese, for example, has both a syllabic and an ideographic reading system, thus housing both reading mechanisms. Some linguistic mechanisms are located in scattered regions of the brain, so language production depends on an adequate process of all the gnostic functions of the brain, among which hearing is the most important of them all. Thought involves language and vision. As far as the blind are concerned, smell, hearing and touch are their vision.

Another way of describing this process is the so-called vocalized thought. The vocalized thought is an important structural function of language that is preceded by initial language learning during childhood and develops along with it concurrently for the rest of life. Vocalized thought can be described as a silent language, confined to the brain regions and specialized in this function that is communicated through speech, writing or sign language. Thus, we have no way of dissociating word, language, vocalized thought and visual thinking, since the word has the associative component of the image.

There can be automatic languages without visual thoughts, however, there cannot be vocalized thought without previously acquired language. So, the thinking process is related to practically all parts of the brain, mainly to memory, language, vision and hearing. The capacity for abstraction is the basis of human thoughts, which is reached at the age of six. At this age, this skill is acquired with the purpose of dealing with intangible elements, such as mathematics.

The word and the capacity for abstraction are found at the threshold of the human cosmos, as they functionally characterize mankind. Language is then considered the most refined human skill. The ability to understand language and to communicate depends on a complex series of interactions of speech centers in the brain. Human language allows us to transcend our experiences. As we give a name to an object, it comes into existence for our consciousness and makes the object that is far from us present. And it includes abstract entities that only exist in our minds, such as actions, states, qualities, beauty, sadness and freedom. The name, or word, that retains in our memory the simple pronouncement of a word automatically represents the object to which it refers in our consciousness, thus forming the mental image of said object associated with a word and its real representation. The act of organizing mental processes, the competence of human beings to use different symbolic vehicles for the expression and communication of meanings distinguishes us from other organisms in nature.

The way in which vocalized thought takes place is very complex. It involves areas of the brain related to language, sounds, writing and speech. These thoughts or ideas before being converted into sound are just a vocalization of the expression of thought. They can be contained or sent to be converted into the necessary movements of the mouth, tongue and vocal cords, initiating the process of speech or hand movements, and manifests itself in writing. When that thought is not sent to the vocal cords and turns into sound, it becomes a vocalized thought — the inner voice, the voice of thought, inner conversation, silent thought, subvocalized thought or the voice of the mind. The main ways to stimulate the natural triggering of this vocalized thought are: reading, listening, visualizing some external fact that evokes a strong emotional reaction and the pure act of imagining the organization of mental processes using the internal voice. The first three are external stimuli and the fourth is an internal cognitive process, based on the content already stored by the individual throughout their life. To better understand what vocalized thought is and how it is conceived, simply do something very common in most people's daily lives: read something. The act of reading with the eyes automatically triggers the beginning of the vocalization process, which is intrinsically linked to language 1. This process is usually transparent and involuntary, and it's difficult for people to pay attention to its execution.

To get an idea of the complexity of the neural network involved from reading to vocalized thought, we're going to follow the process that begins in the so-called fovea. The fovea centralis is one millimeter in diameter and is located in the center of the retina. It has cones and photoreceptor cells that capture light and transform it into electrical impulses, which are decoded by specialized areas of the brain, and have a sufficiently high resolution, around 7 megapixels, to recognize the details of the letters.

We must move our gaze across the page in order to identify at each pause of the eye a word (or two or three). As our eyes are always in motion, the fovea is able to capture several parts of the letters and the brain assembles them in a single image. Our visual system progressively extracts the content of graphemes, syllables, prefixes, suffixes and radicals from words. Finally, two important parallel processing routes enter the scene: the sublexical and the lexical routes. The sublexical route allows you to convert the letters into sounds of the language (the phonemes); the lexical route allows you to access a "mental dictionary" where the meaning of the words distributed by memory is stored in a region called visual word form area, which is systematically activated during reading. This is the final hierarchical stage of extracting visual information in the recognition of letters and words.

---

[1] - There are exceptions, like people with Savant syndrome who read without having to vocalize. Savants see the written page as a photo. They're able to finish a 200-page book in a short time and still remember certain words and passages in the book and on which page the described event occurred. The use of vocalization, in these cases, would delay the interpretation of the information.

This mental dictionary is the final reading process, which activates the vocalized thought or inner voice, voice of the mind. It's a regular and automatic process that is performed in a continuous, transparent and effortless way by the brain of every healthy individual on the planet, for the following purposes: reading, pondering, organizing thoughts, ideas, memories, interpreting messages, among others.

Vocalized thought is also activated by stimulus in the form of mechanical sound waves, the sound. Hearing also activates the process of vocalizing thoughts. A common example: you're in your car and suddenly the radio station plays a song that you like and whose melody was previously recorded in your auditory memory. As soon as the sound waves reach your ears, the sound is processed and you recognize the melody and lyrics. The act of singing the song, either in a low voice or just in your head, is a process of activating vocalized or silent thought.

Another way to activate this mechanism is just thinking without any external visual or auditory stimuli, that is, reflecting internally, in daydreams, or pondering an event that has affected the individual.

We now have the knowledge of what the internal voice is. We started to pay attention to this type of thinking that we usually do not pay attention to, as we never realized that this process could be hacked, violated and exposed.

Figure 1.1 *Visual stimulus. Reading that automatically generates the vocalization of thought; the silent thought in the act of reading.*

Figure 1.2 *Auditory stimulus that automatically generates vocalized thought in the act of remembering a song.*

Figure 1.3 *External and internal stimuli that activate the process of vocalization of thoughts. The human brain is capable of silently thinking about a specific language without the need to express it through speech, writing or gestures. Perhaps it's the only one in the animal kingdom that can do this.*

There is also the second type of thinking that basically depends on the vision and the neural circuits responsible for memory. It's called memory of images or memory of the imagination, which is associated with the visual memory of the word. These are mental images, which depend only on the visual memory used for abstract, contemplative thoughts, and do not require words. They're visual thoughts that can abstract creativity and transcend time and space. As a matter of fact, the human brain is continually creating mental images. This is one of the fundamental ways in which we orient ourselves in the world. The mental structuring of images allows the brain to create relations among objects in the physical space that our senses can detect. Based on these images, we choose how to interact with the world.

Images are the main source of choosing our behavior. There are two key ways in which the mind receives the sense data with which we create these images. One is through what we see; the other is through the language we hear. This is known as a verbal image and has a powerful effect on human behavior. Once a person hears words, the brain immediately processes the sensory data with a coupled image. Whenever we think about something, we evoke a mental image to help create a context based on the reality we live in. Mental images are spatial intelligence centers that take visual perception of the world. To imagine is to work with the image, even with no external visual stimulus, just using visual memory.

To better understand the concept and separate one type of thought from the other — as they are closely connected — let's stop for a moment and try this exercise: close your eyes and imagine a house. It can be a place you've lived in, a residence that recollects a memory of your childhood or another one that is vivid in your memory. And that's it! Visualizing this house in your mind — the images of the object that designates the house — is your brain's way of "virtualizing" the visual reality of the environment around you. It's called visual memory, mental image or imagination image. This memory also makes the visual connection of the symbolism of the word HOUSE, since the sign and the symbolism "house" in no way resemble the object that represents it in our minds nor its representation in the real world. Thus, the meaning of the word uses the same source of visual memory to contextualize its meaning.

Figure 1.4 *Representation of vocalized thought and visual thinking.*

1) Word HOUSE. Phoneme (the smallest unit of sound in speech) is composed of the junction of signs, for example: H + O + U + S + E.

2) The house in the landscape: the real object that represents it.

3) House as it is interpreted in the brain; visual representation of the house in the mind.

4) The feeling combined with the object modulated by the emotional state linked to the recording and the recovery of visual memory, which brings emotions to the forefront and are reflected in the posture of the body.

The decoding of written language is based on oral language. If the auditory/oral language areas are destroyed, we will no longer be able to read normally. We need to keep our hearing intact so that we can hear the vocalization of thought. For this reason, vocalized thought, known as an internal conversation using words, has so much influence on mental processes. It merges at once visual memories, sound memories and abstract feelings, as well as physiological reactions linked to this level of abstraction.

CHAPTER 2 MKTECH
THE ELECTRONIC SYSTEMS THAT FORM THE MIND CONTROL TECHNOLOGY

From now on, all technologies — or modules — that complement each other and form the Mind Control Technology are going to be studied. The acronym MKTECH is going to be used to designate the technology itself. The reason for choosing this acronym will become clearer in chapter 4, volume 2.

Armed with the basic knowledge of how our complex brain creates thoughts, we're going to discover how unscrupulous people and groups devoid of any decency have long been using a set of technology unknown to most people to listen to the thoughts of the population and of specific individuals anywhere on the planet, remotely and non-invasively. Some terms that appear and are repeated throughout the book are going to be discussed. A person who becomes a hostage of this technology and is connected to the system by this weapon is called "Targeted Individual", as well as victim, individual or simply target.

For each chapter that entails a new technology, we're going to use a chart that indicates which technology is going to be addressed within the entire MKTECH universe. This tends to facilitate the visualization of the subject that is going to be highlighted. In subsequent chapters, the technologies already discussed are going to remain visible until we have the complete picture of the entire system.

CHAPTER 2.1

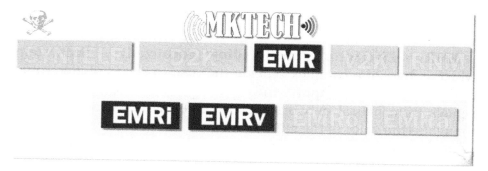

EMR - ELECTRONIC MIND READING

Inside our minds, nerve cells are involved in a "telephone conversation", so to speak. One talks to the other, then the other talks to another one and while the information is transmitted to each neuron that starts to listen to its neighbors, an increasingly complex electrical conversation begins. As the information is transmitted by specialized areas of the brain, it adds more content to the conversation with more and more data until it is demodulated by other areas. These areas interpret the specific electrical signals, giving meaning to this information and making the individual understand what that encoding means.

As an external stimulus, a sound wave that reaches the ears goes through a series of processes until it becomes audible information for the listener — a bird singing, for example. These electrical conversations of neurons in certain key regions of the brain can be amplified, picked up by adjacent antennas and sent back to a remote location where advanced Brain–Computer Interface (BCI) programs dismember these signals — amplified thoughts — into mental images and vocalized thoughts. In this way, thoughts can be decoded and literally heard.

Meet EMR

In order to facilitate understanding, we're going to divide the Electronic Mind Reading into two distinct sources of thoughts, as we've seen in chapter one:

EMRv - Electronic Mind Reading (vocalized) — Subsystem which is part of a complex scheme that uses a series of electronic devices to capture, "amplify" and decode the content of the electrical signals from neural networks responsible for the vocalization of thoughts (the voice of the mind).

EMRi - Electronic Mind Reading (images) — Subsystem which is part of a complex scheme that uses a series of electronic devices to capture, "amplify" and decode the content of the electrical signals from neural networks responsible for mental images or visual memory of thoughts.

Among the technologies involved in the Mind Control Technology, the electronic reading of thoughts can undermine the social, cultural and economic models of modern society, as it is capable of the unthinkable, of making people question whether it's really possible for this type of device to exist today. Unfortunately, the answer is yes! It's already a reality and is being widely used under the table for several purposes, most of them obscure.

All of this happens because this technology completely violates what is the most confidential, sacred and private in human beings; what differentiates mankind from animals: our thoughts. This technology is able to hear vocalized thoughts and see mental images created by the brain. It looks like an episode of a science fiction TV show, but it's the reality to which we're exposed today. Everything that the target thinks is amplified from their brain and captured by a set of electronic equipment, such as radio antennas, microwaves, satellites and adjusted radars, in a remote and non-invasive way. Sophisticated programs translate this neural electrical activity into a human-readable format to the people who will hear and see this thought on a screen using the advanced Brain–Computer Interface. Everything happens in fractions of a second. The individual doesn't even realize that they are having their thoughts stolen.

As we're going to deal with these terms throughout the book, it's worth stressing that people who have their brains captured, kidnapped or connected to these weapons and their thoughts heard by others inside their heads are known as "Targeted Individuals", as well as victims, individuals or simply targets.

The Electronic Mind Reading (EMR) has several practical features, such as espionage, thought hacking, information theft and surveillance. The convenience is considerable for those who are operating the technology. It's possible to connect to it from a remote place thousands of miles away from the target and spend hours, days, weeks, months and even years stealing information directly from the victim's mind, even if the target is physically located in another country.

As this technology is unknown to the general public, the initial impact of having thoughts heard by others — of your cognitive processes becoming public without your consent — twenty-four hours a day and seven days a week is extremely disturbing! The change in the way of dealing with the lives of people affected by this weapon is very clear. Usually, the individual starts to try to control what they will think before they actually think. It seems like a paradox, but it's not. Although this process is extremely difficult and requires extensive training, it's feasible. One could say that this weapon is the embodiment of the violation of all fundamental human rights.

As we discover that it's possible to hear the thoughts and see the mental images of other human beings, we know that we're no longer able to keep any secret. The only place that was believed to be safe is no longer safe, as unauthorized people invade, indiscriminately hacking other people's minds and stealing information and creative thoughts without any physical contact with the target.

The inevitable popularization of this technology will imply a series of consequences that we're not ready to deal with at the moment as individuals. Almost all the foundations of an organized democratic society will crumble when this technology becomes effectively popular. The essence of the technology itself violates all rights guaranteed by the constitution regarding the individuality and privacy of every human being. However, the aggravating factor is the method in which this invasion is carried out. They're using a natural physical phenomenon capable of travelling billions of light years in the universe with no medium necessary to propagate at a constant speed, which is the maximum speed that "something" can travel in the universe, as if there was nothing in the way, such as walls, people, animals, stone or even concrete: the electromagnetic waves. Waves, mostly invisible to humans, are present in everything in the universe, from atomic connections in your body to the way your brain makes neural communication possible, sending and receiving commands to the entire human body. When properly modulated and transmitted, these waves interfere with the electrical functioning of the brain, leaving most people on the planet helpless, since electronic tracking devices are extremely expensive and require advanced knowledge to use them correctly. Most are also inefficient to detect this complex attack.

The bare truth is clear: electromagnetic waves adjusted at a certain frequency and power — in a process of interaction between these waves — are able to interact directly with the brain's bioelectricity. The frequencies used to perpetuate neural amplification and any invasion scheme are the same used by electronic wireless communication equipment on which we are deeply dependent these days: Wi-Fi technology, telephone, internet, radio transmission, television and communication in general.

As it is very difficult to conceive that this type of activity is possible and is happening today, I'm going to enumerate and explain the harmful consequences of this technology when indiscriminately used. There are countless imminent negative consequences of remote reading of others' thoughts, including affecting the sense of self and specific intellectual domains of the people involved. Some events become useless or meaningless, even everyday situations. Moreover, social relationships are deeply affected by it. But don't worry! We're going to see all these aspects in detail throughout this book.

One day, while talking to a good friend of mine, I decided to approach this subject superficially without giving too many details. His reply was: "This kind of technology doesn't exist. If it did, we'd have World War III". This is the natural conclusion that intelligent people come without even knowing the facts and how it all works; others cannot even conceive the idea. On account of that, I'm going to list some everyday situations in which this technology dismantles everything around us. In this way you can see that the technology of hearing people's thoughts and seeing mental images has been around for a long time and has been restricted to military and intelligence agencies, but now it's spreading around the world like a plague, reaching groups of people who are not part of these social circles. New paradigms will be created and will mark a new phase of social relations completely different from what we have today, taking into account several factors that previously did not enter this equation of interpersonal relations in all areas.

Now observe how a commonplace act within the most intimate place of our being causes serious problems when violated. Imagine that you just turned on your cell phone, or your computer, and decided to access your e-mails or your social networks. You sit down, go to the page and the screen displays: login and password. Confident that your data is protected, because in addition to your computer having the most modern firewall and antivirus of the present time, the site has strong encryption, so people will hardly be able to decrypt your data while you navigate the vastness of the internet. Then, you start typing your username and password calmly, without even realizing that even before you mentalize your password and later send the commands to turn the thought into action for the motor cortex to do the complex job of moving your hands and fingers on the computer or cell phone keyboard and entering the password, the initial thought has already been stolen by the hackers' technology at the speed of light, using only the first module of the technologies that make up the MKTECH system: the Electronic Mind Reading (vocalized), also known as EMRv.

Hacking of profiles and passwords from computers, websites and systems

The first systems to collapse will be the authentication mechanisms based on passwords typed by the user, which are precisely the predominant model these days on websites and operating systems. No matter how secure the password, the encryption system or network defenses are, all of this becomes completely useless. The person will always have to think of a sequence of characters to enter the password, and will automatically use the vocalized thought. Test it now. Try to create a new password for some website without voicing your thoughts; try to think of a password without using the internal voice, your silent thought. Practically impossible, right? Even if the person types the password based on the position of the keys on the keyboard, at some point they will try to remember it internally.

The data acquired through social engineering that computer hackers [2] are so dedicated to obtain will be accessible at a glance. Just listen to the target's thoughts and all the data will quickly be available for use without having to face any kind of resistance, thus starting a wave of cybercrime never seen in history.

The solution to deal with this situation will be authentication based on multimodal biometric information, such as your fingerprint, the iris of your eye or voice recognition, or even security tokens and two-step verification, called two-factor authentication with cell phone use to confirm. Not even strong password management and creation programs are spared. In addition to not being accessible to those less familiar with such technology, at some point they would request a master password to open the main password database, which would evoke the thought to be created and typed. So, we fell into the same problem again. Programs that encrypt folders and files that request a password will also become useless, since the password responsible for decrypting the files can be easily accessed in the individual's mind. Creating strong or weak passwords becomes completely irrelevant as well. This is another headache for companies to protect their data, considering that anyone can easily discover the password of an important person on their staff and break into the system on their behalf, steal the database and have access to all the information stored. Of course, you can take a reference file, like a picture, and convert it to hexadecimal format, which

---

[2] - The term "hacker" is going to be used throughout the book to describe the agents and their acts behind the technology. I'm not going into details on whether or not the term should be replaced by "cracker", for example.

would serve as a very strong password. However, this type of procedure is not widespread and it is possible to find out which file serves as a reference by accessing your thoughts.

When I state that any computational structure will collapse, it's not a mere uproar. After all, with the possibility of accessing decrypting keys at all times, encryption and information security lose all meaning, as it violates the principle of data reliability. The protection of information cannot be guaranteed, much less the confidentiality, nor authenticity, since we will never know if the parties really are who they say they are. An old principle by Auguste Kerckhoff — a Dutch linguist and cryptologist who lived in the 19th century—, which is followed to this day will be violated: "The security of a cryptosystem should not depend on keeping an algorithm secret. It depends only on keeping the key secret". As long as passwords have to go through the brain to be generated, maintained or have their origin pointed out, our system will continue to fail.

Authorities that carry out digital forensic investigation can count on this tool to overcome any password on encrypted files. The digital crime investigators headaches will be proportionately relieved as the use of cryptography advances among criminals. Until a final viable solution is found, digital anarchy will be fully established.

Intellectual property theft

There are many unauthorized copies of movies, books and songs on the internet nowadays. This is a major problem that results in numerous revenue losses for those involved in the project. Now imagine a very talented person creating their intellectual property, as they research and work hard for long periods to write a script for a movie, a valuable project, a book or a song. In short, anything that is conceived by society that has value is initially created by thoughts and later stored and cataloged on papers or computers. Before the author — extremely satisfied, hopeful and confident of success — can complete their hard work, while preparing the product for the market, something they came across on the internet makes them freeze in disbelief. It turns out that absolutely everything they created is being sold under the name of a different author who is profiting from their creation. A work that often takes years to complete with enormous added value was patented in the name of another person. They were completely plagiarized and all of their content illegally appropriated.

This is a possibility and is already happening due to the Electronic Mind Reading (EMR). Gradually, people who had their intellectual work plundered will begin to ask themselves: "How did they manage to steal my work before it was finished? Did they break into my computer?". The answer is no! Hackers only connect the EMR to a person of interest and leave it twenty-four hours a day and seven days a week, copying all the thoughts created, including the intellectual means that made the hacked author reach a certain conclusion in their book, script or project of any kind. They also steal the creative cognitive process and product content.

Brace yourselves, because intellectual property is in serious danger. Take a good look at the seriousness of this new type of theft: people can illegally download a book, a movie or a song on the internet without paying anything — the famous piracy—, but even with financial losses the product will continue to belong to the real authors, studios or band that composed the song, for example, which is completely different from illegally taking over it and releasing it as if the product had been written by the thief, before it was even completed by the real author. Think about what will happen to the author of a book, the creator of a series involving hundreds of millions of dollars, the entire industry and people who orbit and depend on this content to survive. All of this will collapse as well.

Nevertheless, the most serious situation would occur with a newcomer without sufficient resources, such as a promising producer, author, composer or programmer starting their career by creating great material with the potential to stand out among the great. If the material were stolen straight from the mind, patented and published by the mind hacking scheme, the author would have extreme difficulty in proving by legal means that they were the original author of the idea; that they had been a victim of theft of thoughts [3].

---

[3] - Many victims of theft and psycho-electronic torture have unsuccessfully tried to contact the authorities. They were treated as mentally ill, were admitted to psychiatric institutions or were completely ignored. We're going to better address this topic in the following chapters.

Theft of Confidential Information

Secretive technological projects that involve expensive research and investors — in other words, a lot of money — are always the most targeted, whether they are a new medicine, system, application, weapon, concept or a patent. The biggest fear of companies that develop new and expensive products, including the military, is undoubtedly industrial espionage. By using the Electronic Mind Reading, it is possible to follow day by day the development of the new product and discover absolutely everything about it, thus causing enormous losses for the company. It's even possible to understand how the hacked-minded human developer responsible for the product development uses their creativity to reach the final product.

Even a large enterprise that believes it is safe by shielding the entire headquarters building is not out of danger. Shielding the building — which is extremely expensive to implement — would block electromagnetic waves from penetrating its interior and prevent the theft of thoughts of its key employees. However, the problem would only be partially solved. As soon as that employee leaves the safe place of work and is heading home, for example, his or her brain is exposed to remote amplification (reradiation) and thoughts may be stolen. Invariably, this employee will think about how to solve a work-related problem, be it on the way home, during a walk in the park, having lunch in a restaurant, exercising in a gym or during a moment of reflection. This act of naturally diverting the flow of thought during activities that have nothing to do with work — as reported above — comprises the principles that govern the thoughts that make us organize reality, select and privilege certain data and eliminate or subordinate others, leading us to think about our work and the object desired by thieves.

Even few details give clues to what is being created. Fragments of thoughts are enough for the purpose of successful industrial espionage. Mind hackers can easily steal the entire project, as they listen to the target in their spare time for months. That is why the dissemination of this technology is crucial, so that people continue to feel free to think, to let the natural cognitive processes flow and to not have to worry about hiding the thought on any given subject. The remote violation of thoughts by third parties is extremely serious and sets a precedent that we're not ready to deal with at the moment.

Theft of credit cards and PINs

You're at home and are about to shop at your favorite online shopping website. It's a secure site that hardly has any problems with internet payment transactions. You're getting ready to enter your credit card details, but as soon as you start the process of reading the data in silence (such as the numbers, the card security code or the expiration date) all data will be stolen straight from your head even before you enter the information on the website. Once the numbers are read, you vocalize the thought — an inner voice telling your brain what your eyes are reading. The thought is then captured, amplified and sent to the mind hackers' computers. This is yet another model that will collapse soon and will make the use of a credit card based on security codes, as we know it, unfeasible.

ATM withdrawals without biometric fingerprint technology

There are some places in the world that in addition to requiring your bank PIN to access important services (i.e., financial transactions) you must have an access code (which is made up of letters). So, in these places, the security of ATMs that do not use biometric systems is still based on passwords stored in people's minds — which are accessed through vocalization of thought—, by means of an access code based on the association of each side key of the ATM to a list of five different syllables. This authentication uses set theory to create a different subset of possible codes for each customer with 25 letters or syllables. As soon as you think of the association of letters to choose at the ATM screen, the information will be easily stolen by this technology operators.

To understand the size of the problem we're facing, I propose a challenge: the next time you are at an ATM to withdraw money, try to outwit thought thieves. Let's see how to do this:

If your hypothetical combination is [po-ta-to], try to vocalize it inversely, randomly or completely different from the actual syllables. When the set of letters appears and you're looking at the screen to enter "**po-ta-to**" try, first of all, not to think or vocalize the correct combination before starting the process of choosing a syllable. A hard task to achieve the first time, right? It requires a huge mental effort not to search the memory and transform this data into vocalized thought, reversing the way the brain normally works. Regardless, each code will appear in five sets with mixed syllables. So, for each screen, choose the correct ones, however, *think of others*. For example: if your first syllable is [**po**], locate it on the screen, but think of [**to**]. Although [ta] is the second correct option, you should think of [**ma**]; finally, instead of [**to**], think of [**lo**]. Try it with your access code combination at an ATM and see if you can trick the mind hackers.

That's a mission almost impossible to be performed. It's extremely complex and difficult, as our mind is not used to this type of trick: reading the letters on the screen, remembering the correct ones, searching for them among the options and entering them silently as you think of others. This requires years of training. Unfortunately, this combination-based security is effective if someone is trying to see what you're typing and in the case of hidden cameras intended to record PINs and access codes. Nonetheless, it's useless for thought thieves, since the sequence will invariably be vocalized before being chose on the screen or on the ATM keypad. In other words, in just one day you can capture all combinations, PINs and account data of everyone who withdraws money or accesses their bank accounts at a single ATM. Hackers don't even have to go near the ATM or break inside. In the end, the bank's security completely collapses.

End of interpersonal relationships

Take an everyday discussion at work or at home as an example. The thought modified by primitive emotions, such as momentary anger, ready to be expressed in the face of a common and relatively frequent event in everyone's life. In the heat of the moment, you'd speak your mind; you'd vent to the person with whom you discuss everything you don't like about them, everything that is kept inside you — a reflection of the problems of everyday life in society. However, this person is your boss, your wife or your husband and it wouldn't be prudent to express yourself that way. Self-control would take place and wouldn't put these thoughts into words, as you could be fired, a relationship could end or even worse! Human behavior must be governed by reason and emotion. Placing too much emphasis on any of the two entities alters the personality. Knowing how to manage reason and emotion is a constant internal struggle in people's lives. But the arrival of EMR changes everything.

The processes that create these thoughts stored in your mind just for you, totally free to move among your brain areas where you can vent about someone else; everything that is embedded in your strictly private cognitive framework can now be amplified and heard, making it public! Imagine what would happen if both individuals heard such thoughts? Interpersonal relationships in all spheres of life would end. This would create unnecessary confrontations that'd have previously been avoided, in addition to causing severe distress, as some of the thoughts are associative and we don't control the shortcuts the brain creates to facilitate cognitive processing. These processes exposed in public may be interpreted in a pejorative or unfriendly way, and they can profoundly affect the personality, causing severe states of inner discomfort.

End of privacy

By using the Electronic Mind Reading (EMR), an individual can hear the thoughts of all the people around them in a short time. Human curiosity about the private life of others is the trigger for the indiscriminate use of this technology at any time. This will soon be one of the reasons for its popularization and miniaturization. Please bear in mind that this isn't the same as spying on someone using binoculars or listening to a third-party phone conversation. This technology violates the cognitive processes that define mankind, the product of the mind, our rational thoughts. That is why it's so dangerous and puts all civilized democratic processes at risk!

End of the right to free thinking

Now, imagine having to think before you think; thinking if you access a certain memory, if you finish expressing a thought or if you think about the past. With this technology at work, in addition to having to deal with the daily and intellectual activities inherent to life in society, we'll have to think before we actually realize the thought. It's a complex and difficult process to understand for those who weren't targeted or didn't have their brain connected to this technology, thus setting a new precedent in history: the inspection and publication of everyone's private thoughts. It may become a weapon of ideological retaliation and inspection, or of restraint to any type of internal cognitive process.

We'll be judged by others, our thoughts will be disclosed even before they become action and are presented in the form of behavior, which would be the last natural action. A very serious fact that leads to numerous philosophical and social considerations that must be faced, including in terms of legal consequences, since the Judicial and Legislative Branches will have to start deliberating on this subject. Knowing the existence of thought stealing directly from people's brains, where would the privacy frontier be? Are thoughts public? When people are at home, are the walls of their houses and everything contained therein, in fact, private property? These are new 21st century dilemmas that we'll inevitably have to face as a society. I could spend dozens of pages listing various situations in everyday life that seem innocent at first sight, but whose context is radically changed with the use of this weapon. As the book progresses, we're going to understand better what is to come.

You can already understand the devastating consequences of this technology in this small initial sample. An authentic weapon of the 21st century, modern electromagnetic weapons that were built to an end: to interact at a distance with the electrical functioning of the human brain, deforming, altering and stealing the content modulated in this activity, the complex thinking and the consequences of its functioning, including the corners in which the subjective parts operate, such as logical and abstract reasoning. It's important to bear in mind that the EMRvi – Electronic Mind Reading (vocalized/images) is just a fragment, an important subsystem, a module of the MKTECH (Mind Control Technology) system.

As we go deeper into the reading, we're going to see the functioning of each subsystem and its characteristics, as well as the peculiarities of its performance in the minds of mammals, acting as an integrated system in which reality becomes stranger than fiction. Then, the next natural question would be: *but how does it really happen?* How is it possible to amplify my mind's thoughts in a remote and non-invasive way?

2.1.1 - How does the whole process of amplifying brain waves and extracting their contents from a remote position take place?

Electromagnetic Waves and the Brain [~~^~~~^ {^~~^~~~^].

Electromagnetic phenomena have been known to mankind since ancient times. However, visible light was the only known part of the electromagnetic spectrum for a long time. The Greeks were already aware that light traveled in a straight line. They even studied some of its properties, which are part of what we now call geometrical optics. The development of events related to the discovery and the understanding of electromagnetism occurred slowly over the centuries. They were developed by a number of prominent personalities who contributed to achieving a better modern understanding of one of nature's most important phenomena. In the following, I'm going to make a brief historical summary of this study.

* Around 600 BC, the discovery of the electrostatic charge was described by the philosopher Thales of Miletus, who rubbed an amber with an animal's skin and verified that this amber attracted pieces of straw after the friction. In the year 900, a shepherd named Magnes discovered that lodestones, naturally occurring magnets, had the ability to attract iron. However, qualitative and scientific studies of electromagnetic phenomena only began in the 17th and 18th centuries.

* As early as 1750, Benjamin Franklin discovered that lightning is electrical discharges and invented the lightning rod to attract them. Soon after, in 1785, the forces between stationary electrical charges were explained by the laws of the Frenchman Charles-Augustin de Coulomb. Also, the electrostatic and magnetostatic fields — fields that don't vary with time — were formulated mathematically. They named the electric charge unit after him.

* The study of the relationship between electric and magnetic fields and the behavior of time-varying fields had its early progress throughout the 19th

century. In 1800, Alessandro Volta, an Italian physicist, invented a generator that converted chemical energy into electrical energy called voltaic pile. The electric potential unit was named Volt in his honor.

* The first discovery of electromagnetic radiation other than visible light came in 1800, as William Herschel discovered infrared radiation. In his experiment, Herschel moved a thermometer through light split by a prism, decomposing it, then measured the temperature of each color. He noticed that the temperature rose from purple to red, and that the highest temperature was beyond red, in a region where no sunlight was visible.

* In 1820, Hans Christian Ørsted observed that electric currents created magnetic fields that caused the compass needle to deflect. The theoretical explanation would come with André-Marie Ampère, who was one of the founders of the Electromagnetism Theory. Therefore, the unit of electric current, the ampere, was named in his honor.

* In 1827, Georg Simon Ohm formulated the famous Ohm's law, which connected the electromotive force with the resistance and the current.

* In 1830, Joseph Henry discovered the electromagnetic phenomenon of self-inductance and the unit of inductance, in the International System of Units (SI), was named in his honor.

* In 1831, Michael Faraday verified that a time-varying magnetic field induces an electric field (explaining Joseph Henry's discovery) and introduced the concept of lines of force. So, the farad, the SI derived unit of electrical capacitance, was named after the physicist. Werner von Siemens, a German inventor and industrialist responsible for several inventions, such as the pointer telegraph, the electric elevator, the selenium photometer, the electric generator and the dynamo, was honored with the S symbol that measures the electrical conductance.

* In 1835, Samuel Morse invented the first telegraph and co-developed the Morse code, which started to be used in telegraph transmissions.

* The Joule unit was named after the English physicist, James Prescott Joule, who demonstrated the equivalence between heat and mechanical work in 1849. The experiment involved the use of a falling weight to spin a paddle wheel in an insulated barrel of water which increased the temperature.

* James Clerk Maxwell, a notorious Scottish physicist, formulated the mathematical foundations for the analysis known today as Maxwell's equations, a very important landmark of 19th century mathematics and physics, which unified Optics and Electromagnetism by demonstrating that light is constituted by electromagnetic waves and propagates in the air at a speed of about 299 792 458 m/s. He was the pioneer in the practical demonstration that it's possible to obtain magnetic fields from the oscillation of certain electric charges, which generate electric fields.

New magnetic fields originate from the variation of these electrical flows, and this interaction results in the occurrence of electromagnetic waves. Inspired by the laws of Coulomb, Faraday and Ampere, Maxwell was aware of the existence of electromagnetic waves. This initial perception, however, was only proven years later by the German Heinrich Hertz, in 1888, which started the era of telecommunications. The unit of frequency was named the "hertz" in his honor. It expresses, in terms of cycles per second, the frequency of a periodic event, oscillations (vibrations) per second.

Frequency is the number of occurrences of a repeating event per unit of time, creating a well-defined pattern that everyone can use and synchronize with each other: the hertz (Hz). One hertz means "one cycle per second"; 100 Hz means, "one hundred cycles per second", and so on. The unit may be applied to any periodic event and is used in sound frequency and electromagnetic radiation, for example.

Through the unification of electrical and magnetic phenomena and their understanding, it was concluded that electromagnetic radiation and light were the same. Therefore, the definition was: electromagnetic waves consist of an electric and a magnetic field like a transverse wave that vibrates perpendicular to the direction of the wave or path of propagation. The two fields vibrate at right angles and with each contraction and expansion a field is created in the opposite direction. The vibration of the electric field creates a magnetic field and the vibration of the magnetic field creates an electric field. The result is a wave that travels at the speed of light and requires no means to propagate, unlike mechanical waves such as sound, which need some medium such as gas (air), liquid (water) or solid (steal). That is why it's the only [4] wave that propagates in the vacuum of space.

---

[4] - On September 14, 2015, the Laser Interferometer Gravitational-Wave Observatory (LIGO) first detected gravitational waves generated by the collision of two black holes, confirming their existence and causing deformations in the fabric of space-time to the extent of 1% of an atom's diameter. This wave is not the focus of the book.

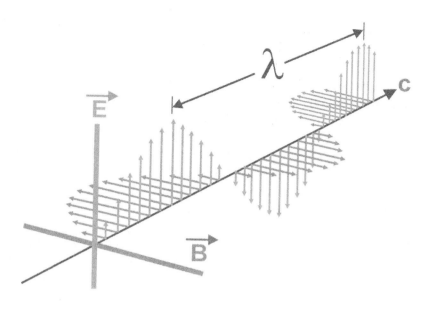

Figure 2.1 *B - Axis of the magnetic field and E - Axis of the electric field. C - Direction of wave propagation. λ (lambda) is the length of the wave. When it moves fast, at the speed of light, the released energy seems to take the form of waves.*

Electromagnetic waves are present in the daily lives of all of us. They're part of our modern society. As wireless technology advances, although most of the spectrum is invisible to us humans, we can find them all around us: on AM and FM radio broadcasts, in your car keys, routers, television broadcasts, cell phone communication technology, home microwave ovens, X-rays in hospitals and airports, conversations between control towers and aircrafts, modern drones, satellites that orbit the Earth, and so on. Even the light in the visible spectra and solar rays are electromagnetic waves.

Depending on the wavelength, the seven types that make up the electromagnetic spectrum are covered, ranging from radio waves to microwaves, infrared, (visible) light, ultraviolet, X-rays and gamma rays. Any disturbance in a current or in a distribution of charges spreads through space as a wave, whose speed is the same as that of light in a vacuum. The speed of light in a vacuum is constant for any observer, regardless of the movement of the observing source, so the limit speed of the universe is not infinite. Electricity and magnetism are closely linked together.

The complex Quantum Mechanics is also, in essence, based on studies on the relationship of light and matter. In general, the various types of electromagnetic waves differ in terms of wavelength, a fact that changes the frequency value, and the way in which they are produced and captured. That is, from which source the electromagnetic waves originate and which instruments are used so that they can be captured. However, they all have the same speed, in a vacuum, of about 3.0×108 m/s or 300,000,000 m/s. Its exact value is defined as 299,792,458 meters per second and can be originated from the acceleration of electrical charges. X-rays and gamma rays are absorbed by the atmosphere, but spread freely through space.

Despite this, in some determined frequencies, it's known that electromagnetic waves have the ability to interact with molecules of living organisms through resonance. If both frequencies are equal, the molecules capture the oscillation and vibrate at the same frequency. On the other hand, the interaction of the microwave oven with the water, which is heated due to the agitation of the particles[5], occurs at the frequency of 2.45 GHz — between 700W to 900W.

---

[5] - The water molecules act as an electric dipole and are affected by variations in electrical charges of the microwave that cause the constant rotation of the particles. This process is called dielectric heating, which is different from resonance.

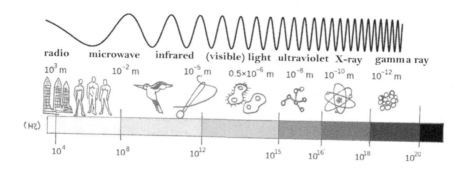

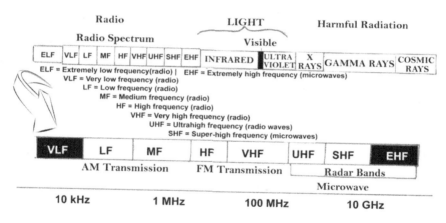

Figure 2.2 *Electromagnetic spectrum.*

The discovery of everything that involves electromagnetism and photon properties was one of the greatest humanity's findings. It was through this discovery process that we were able to understand almost everything we know about the scientific observation of the universe. We also managed to know how bright the stars, galaxies and planets are, even the most distant ones, as well as their temperature and speed in space, their rotation, the chemical composition of the materials that form the atmosphere of these planets, even those located outside our solar system, the exoplanets.

Modern science, astronomy, medicine, communications, television, computers, cell phones, basically everything that exists today exist thanks to electromagnetic theory. Without it, it would be impossible to test models of the cosmos, for example, and astronomy as we know it wouldn't exist. In the long run, our understanding of everything'd be extremely limited.

2.1.2 - Brain Functioning and Electromagnetism

The brain is also no different. We're made of the same materials available in the universe and we operate under the same laws of physics that govern nature. Thus, our brain is a formidable antenna that receives and transmits electromagnetic waves, since cells like the neuron use electrical impulses to communicate by means of axons, which are similar to the operation of the telegraph, sending electrical information to distant locations in the nervous system.

Among the countless interactions between waves and living organisms, the most noticeable is that of light, which is absorbed by the cells of the eye causing visual sensation. This interaction occurs in the frequency of the visible light in the electromagnetic spectrum. The retina, a light-sensitive layer of tissue of the eye, specializes in capturing the light energy that comes from the environment and transforming it into bioelectric potentials, in order to encode in them the information contained in the incident stimuli, which later forms the vision. The UV spectrum — Ultraviolet — is found at a higher frequency of waves, above the visible light range. UVA rays cause an immediate tanning effect by oxidizing the melanin, whereas UVB rays hit cholesterol in the skin cells, providing the energy for vitamin D synthesis to occur. In a spectrum below the visible light is the infrared radiation, which can be absorbed by the skin and felt by specific areas located in the Neocortex via thermoreceptors responsible for the sensation of heat and cold. Radio waves and microwaves are found in the spectrum just below the infrared. If properly modulated at the correct frequency, and at a determined intensity, they can directly interfere with any human brain, altering its functioning in almost all cortical areas: from hearing, vision, sleep, dreams, feelings and emotions to the content of thoughts and much more, as we're going to see throughout the book.

The primary cortical areas — that comprise the motor and sensory regions — capture external stimuli such as electromagnetic and mechanical sound waves. As soon as they recognize the stimulus, they analyze and decode the outer model, synthesizing the stimulus from the transduction [6] in order that the outer world corresponds to an inner world in the individual's brain. For all of this to happen and be perceived, it's necessary to go through the process of recognizing objects through one of the senses, such as the visual and the auditory ones. It appears, therefore, that the most important external stimuli received by the organism are comprised of electromagnetic waves.

It's worth knowing that the complexity of the brain surpasses that of any existing computer today. There are approximately 100 billion nerve cells or neurons in the human brain and 100 trillion contact points, called synapses. It's an integrative system that works through cooperation and interaction between different areas, forming a unit based on an organized series of bioelectrochemical reactions between nerve cells. Each of these cells carries a negative electric charge inside itself and a positive charge along the outside of the cell membrane. This creates a small electric battery that works using electrical discharges generated by exchanging ionized substances [7] for neural communication. Axons make contact with other cells at these junctions (synapses) and create the communication circuit for each neuron.

Neuromodulator molecules are responsible for transmitting information previously encoded by electrical signals to the neuron. Neurons, in turn, continuously receive impulses at their dendrite synapses from thousands of other cells that are responsible for conducting these impulses. The stimulus is initially coded in the amplitude of the electrical responses of the sensory receptor. From there, from neuron to neuron, the code alternates between the occurrence frequency of the nerve impulse along the nerve fiber, passing the impulse to the next neuron. In this way, the information is conducted by neural circuits that comprise neurons located deeper and deeper

---

[6] - Sensory transduction is the process by which the stimulus from the environment activates a receptor and this stimulus is converted into electrical energy to be sent to the Central Nervous System (CNS).

[7] - For electrical synthesis to exist, it's necessary to exchange ionized substances between cells and the environment. The inside of the resting cells is negatively charged compared to the outside of the cells due to the difference in ion concentration. Sodium (Na +), Calcium (Ca2 +) and chloride (Cl) are found in the extracellular space, while K+ is concentrated in the intracellular space.

until the cerebral representation of the sensory event is formed. Thus, internal processes are also subject to the laws governing electromagnetism, since all communications among cells are electrical, generating weak diffuse magnetic fields in all brain and motor activity in the body.

To locate and visualize our biological computer, let's take into account the physical disposition of the brain areas according to modern concept, which is divided into five main areas. Each area is called *lobe*.

* **Frontal Lobe** - the frontal lobes influence the motor activity learned and the planning and organization of expressive behavior, a place in which conscious and rational thinking are processed, and synchronize activities.

* **Parietal Lobe** - the parietal lobe is located between the frontal and occipital lobe and above the temporal lobe on each cerebral hemisphere. It's involved in functions such as the sensation of touch, cognition, spatial orientation, visual perception, etc. Many lateral areas provide precise visual spatial relations, but they also integrate these relative perceptions with other sensations to create a personal awareness of the trajectory of moving objects. The knowledge of body parts is also generated in this area. In the dominant hemisphere, the inferior parietal area coordinates the mathematical function and is closely related to the recognition of language and vocabulary memory.

* **Temporal Lobe** - the temporal lobes are involved in visual recognition, auditory processing, memory and emotion.

* **Occipital Lobe** - located in the rearmost portion of the skull, the occipital lobes are part of the posterior telencephalon. This lobe's function is almost exclusively visual. The primary visual cortex [8] is contained in the walls of the calcarine sulcus and some of the surrounding cortex. It's a visual cortex of association — represented by the remaining occipital lobe, also extending to the temporal lobe — which reflects the importance of vision for the human species. It's involved in the processing of visual information.

* **Insular Lobe** - the insular lobe (also known as insula and insular cortex) has a triangular shape and is a part of the cerebral cortex located in both hemispheres. It's surrounded by the anterior, superior and medial limiting sulci, and is divided into two parts: the anterior insula and the posterior insula. It's believed to be involved in diverse functions usually linked to emotion, including the sense of taste.

---

[8] - Cortex, Latin term for "bark". Gray substance in which the cell body is found. It's called cortex when they're arranged in parallel layers.

Speaking, reading and writing

As we've seen in the previous chapter, there are two different types of thoughts that are captured by the MKTECH system: the vocalized one and the visual one. The vocalized thought involves words. It's the internal conversation, silent conversation or voice of the mind. It's the operation that takes place before becoming concrete actions of speech or writing, our interiorization and symbolization, the voice of reading. The visual thinking (mental images), on the other hand, is linked to visual memory, as well as the visual cortex. However, it doesn't need language and it's the most important thought regarding creativity and abstraction. But what happens physically in the brain when we think in a vocalized or visual way?

The speech production and the articulation of words, which are the raw material for vocalized thought, depend on a complex series of interaction between speech centers in the brain. Therefore, language recruits more distinct areas of the brain to be performed, understood and formed. Despite using a complex neural network, which involves practically all parts of the brain, such as the thalamus, the main areas involved in vocalized thought are Broca's area, responsible for speech production, and Wernicke's, responsible for comprehending and producing the meaning of the word.

It works like this to convert thoughts into sounds: when you read a word or hear someone speaking it, the mental lexicon in Wernicke's area recognizes the word and interprets it according to context. Then, in order to pronounce the word, the information must be transmitted from Wernicke's to Broca's area via the arcuate fasciculus, a bundle of axons (nerve fibers). This information is then routed to the motor cortex, so that you can pronounce the word. This model is based on the anatomical location of brain areas that have distinct functions.

Capture of subvocalized, vocalized or silent thoughts until they actually turn into speech:

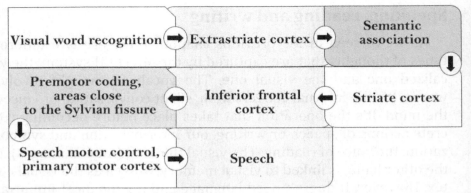

Figure 2.3 *Process of transforming silent thought into speech.*

Vocalization is the expression of thought into recognizable words that make sense internally to the person, even if they remain silently in the individual's mind or become an expression of behavior such as speech. This process occurs one step before signals are sent to the vocal cords. The speech control center, as we communicate, tells the mouth what to say. Thoughts sent to our vocal cords are converted into speech and, at that moment, the signal is amplified, captured and retransmitted to auxiliary antennas. So, the raw signals processed — which would later be sent to the vocal cords—, reach the computers for decoding with the content of the vocalized thought already "embedded" in them. All the thieves have to do is to extract them from "inside" the wave that carried the content there. It's worth mentioning that this is one of the methods to capture the vocalization of thought. There is a more effective one, which we're going to see later.

Mental images, on the other hand, are created using basically our visual memories. Even by mapping all different representations of the visual fields of the brain, identifying all cortical areas involved in the various forms of memory and visualizing the active areas during the imagination, it'd be unlikely to gather this information, since the circuits are fragmented throughout the encephalon. Any type of information processed by the brain involves the highly distributed recruitment of neuron populations. For example: inferior temporal neurons participate in the coding of visual attributes, serving as integrators of information about depth, color, size and shape registered in the prestriate cortex. Therefore, there are several areas working together. It'd be necessary a place where the signal travels before it is scattered by the whole brain in order to capture the thought signal, the image already processed or preprocessed. That place actually exists, both for the vocalized silent thoughts and for those of the imagination (the visual ones), and is called Thalamus. The signal amplification acts on it. However, in order not to be boring for the reader who is having the first contact with the subject, we're going to study this area later (in chapter 2.4) when we're more acquainted with the dynamics of all processes.

That said, operators who steal the victim's thoughts use a range of electronic equipment, a giant infrastructure with the help of satellites designed solely for this purpose. Thereupon we are going to gradually progress until chapter 5, volume 2, in which we're going to better understand everything behind this technology. The purpose of this chapter is focused in demonstrating how easily the brain is hacked from a distance.

We just stop thinking when electrical activities cease — when we actually die or have brain death. Taking advantage of this innate characteristic of the brain, EMR sends an uninterrupted signal. The wave of this signal is added to the wave already demodulated and interpreted by the areas of the brain involved in the constant flow of thought. As soon as the image in your mind is formed — as soon as the house you imagined in the previous chapter turns into a recognizable image of a house in your mind —, the neural network previously mapped by electronic equipment that emits electromagnetic waves is amplified, captured and modulated [9] by a receiver in which a computer quantifies the signal and transforms it into digital format. In this location, it's remodulated and retransmitted in encrypted form and compressed to another receiving antenna until it reaches the remote base, in which the information is deciphered, decompressed and the received data is interpreted into sounds, images

---

[9] - Modulation is the process in which the information transmitted in a communication is added to electromagnetic waves. The transmitter adds the information in a special wave in such a way that it can be recovered in the other part through a reverse process called demodulation.

and writing [10]. This is how the thought-reading operation takes place.

The remote base is the place where the information contained in the wave is demodulated, decompressed and deciphered by a series of computer programs that are part of the support system of the Mind Control Technology and start to interpret the data, thus processing the images of the Targeted Individual's thought on a screen. This occurs almost in real time given the properties of the speed of communication in a constant and uninterrupted manner.

Remember: electromagnetic waves travel at the speed of light **299,792,458 m/s**. Light, at that speed, could circle Earth more than seven times in one second. It's the same transmission speed as microwaves, cell phones, TVs and radio waves. It's impressive how quickly the thoughts of the target are carried "through the air" and travel great distances until they're heard by the system operators.

The brain has no natural defense against electromagnetic weapons and will never have it. So, it ends up interpreting the invasive external waves as if they were part of the transduction system. This method of stealing thoughts practically finds no obstacles nowadays, due to the ignorance of the technology in question and the nature of light. The current techniques for remote amplification of brain waves and the content embedded in the waves — modulating signal — are very advanced and extremely complex, varying according to the program used to capture thoughts. However, the principle is always the same with little or no variation:

---

[10] - Writing occurs in a similar way to word recognition programs that transform it, in real time, into texts called automatic transcription — similar to those on YouTube, or in your smartphone's translation program.

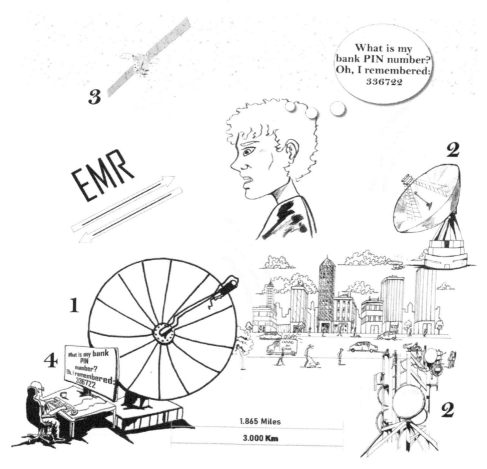

Figure 2.3 *EMR. Amplification scheme of vocalized thought and visual thinking.*

1) Internal antenna that communicates with more powerful external antennas, responsible for sending signals and amplifying them in the victim's head using complex frequency combinations, signal cancellations and reradiation.

2) External radio or microwave antennas that emit and receive the signals resulting from this amplification.

3) Satellites relay the data captured over great distances; they also serve to locate the targeted individual anywhere in the world. Moreover, they actively participate in the process of amplifying thoughts.

4) Devices that decode all information received back to the internal antenna. The received signals are interpreted by an advanced system of a Brain–Computer Interface that displays the result of the target's captured thoughts on the thief's screen.

* **Please note:** – The distance from the target to the base where the initial attack begins and which receives the feedback signal with amplified thoughts is more than 1,865 miles in this example.

* **Please note as well:** – The illustrations don't reflect the actual proportion of the objects. They're designed to highlight certain devices or events. This concept is going to continue throughout the book.

Summing up

From a remote location, several electromagnetic signals of different frequencies are transmitted simultaneously to the subject's brain in such a way that the signals interfere with each other to produce a wave configuration that is modulated by the bioelectric waves of the individual's brain. The resulting interference wave, which represents brain wave activity, is amplified and retransmitted from the brain to a receiver where it is demodulated. **This phenomenon is also known as reradiation.** The content of **the thought is re-radiated** containing the captured information modulated in its signal. The content is then sent to a Brain–Computer Interface (BCI) for processing and analysis. The demodulated wave can also be used to produce a compensating signal, which is transmitted back to the brain to effect changes in the target's electrical activity and flow of thought.

The electromagnetic waves of the MKTECH system cross most urban obstacles. To find out if you're exposed and vulnerable to attack, we can take the signal receiving antenna of your cell phone as a parameter. If from where you're, the antenna is able to capture any type of frequency — for instance, 3G, 4G, 5G, FM, AM and TV — then you're for sure subject to EMR attack. Even in places that these transmissions don't reach, such as subways, planes at high altitudes, tunnels, oceans and isolated forests, the system continues to access the individual's brain. Depending on the type of criminal group and its apparatus and infrastructure — satellites, antennas, radars, organizational capacity and purchasing power —, the victim will be consistently and continuously monitored anywhere in the world. So, it only takes a set of signals of the same phase and of greater power to interact with the brain's electric field, so that the interference gains enough strength and is captured and remodeled, reaching the nearest receiving medium.

The computer program that demodulates the conversation content of the amplified neural circuits in the formation of thought is the equivalent of an interpreter or compiler that has characteristics similar to the specialized cortices responsible for transforming this bioelectric conversation into human-readable information. That is, a highly sophisticated BCI. As soon as the information is received, the program is able to interpret, decode and transform the data into images, sounds and writing, as specialized cortices would do in the brain.

The amplification of thought occurs via the superposition of waves. The superposition is also called interference in some cases. It's the phenomenon that occurs when two or more waves meet, generating a resulting wave equal to the algebraic sum of the disturbances of each wave. Therefore, by the principle of superposition, the presence of a third charge doesn't alter the force between the other two, however, it adds its own contribution vectorially. This type of superposition is called constructive interference, so that the resulting wave has a higher amplitude than the previous waves.

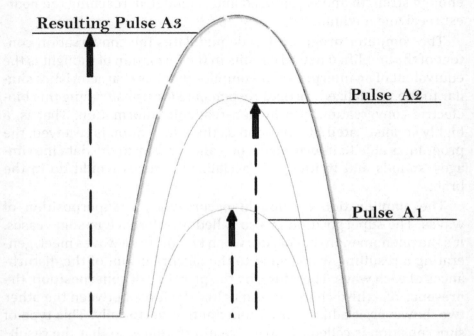

Figure 2.4 *Example of constructive interference, the physical phenomenon that amplifies the brain's original signals to be picked up by antennas and redistributed by a new carrier signal, creating a modulation.*

Out of Phase

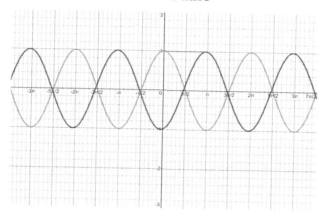

Approaching Phase

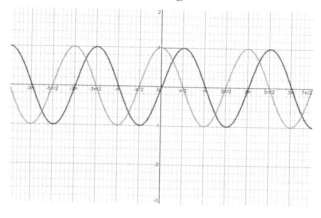

In Phase

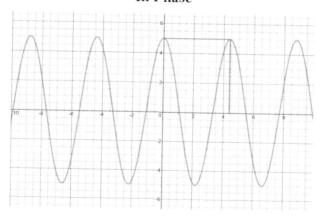

Figure 2.5 *After a complex process of interaction between them, the signals that start at radio frequencies (3 kHz) and go up to microwaves (300 GHz), and as they come across the parameters of the electrical rhythms of certain cortical areas of the target previously mapped in neural biometry, will result in signals capable of interacting with the low electrical frequencies of the brain, "amplifying" and reradiating them, including vocalized thoughts and visual thinking when traveling through their specific thalamic pathways (LGN – Lateral Geniculate Nucleus and MGN – Medial Geniculate Nucleus). The electrical rhythms and signals can vary from ULF (Ultra low frequency) of 4Hz to ELF (Extremely low frequency) of 40Hz.*

The Electronic Mind Reading discussed in this chapter is just a module that makes up a complete system of these 21st century psychotronic [11]/neural weapons, which use advanced methods of brain electromagnetic interference in a remote way. In the next chapter, the reason the technology is called a "weapon" is going to be clearer.

This is then how the major thought extraction technique takes place. The content can be easily hacked directly from the mind of anyone on the planet. You just need to have the appropriate hardware and software equipment for that purpose. However, there is still a lot to understand on this subject and, as we go deeper, everything is going to make sense. How to remotely enter data into the brain is going to be covered in the next chapter.

To further illustrate how this type of attack works and to exemplify it with something from our reality, let's make an unusual analogy using the old vinyl record. Yes, the phonograph disc record! Some of you will remember the old vinyl, while others may never have had contact with this type of media that was very popular before the CD's arrival in the mid-1990s.

2.1.3 - The vinyl record

How does vinyl work? How is sound recorded? How does vinyl reproduce sound and what is its relation to the theft of vocalized thoughts?

The vinyl record, also known as phonograph disc record, long-playing vinyl record (abbreviated: LP), or vinyl for short, is an analog

---

[11] - Psychotronics is a modern term created to refer to new electromagnetic "non-lethal" weapons, which interfere with the electrical functioning of the brain, disabling the functioning of the mind, hacking, debilitating and depressing. "Psycho" from psychological, mental; "tronics" from electronic, which generates the electromagnetic wave.

sound storage medium in the form of a flat disc, manufactured using electromechanical processes and made of a plastic material called polyvinyl chloride (abbreviated: PVC, colloquial: vinyl). It was invented by engineer Peter Carl Goldmark to replace the old 78 rpm discs used for audio storage since June 21, 1948.

A vinyl record is essentially the physical representation of "sound waves" pressed and imprinted into a vinyl surface (PVC or acetate). In other words, the grooves represent the sound waves as we hear when playing it on a record player.

Nowadays for vinyl to be recorded it's necessary to follow the following process:

1) The music is electronically mixed in its digital form.
2) The disc rotates at the speed of rotation.
3) The music that is played on the computer is transferred to a needle that carves the tracks according to the frequencies and vibrations of the sound wave shape with its valleys and peaks.

For the music to be played, the reverse process takes place: the tracks sculpted in the sound format are traversed by a needle that vibrates according to the frequencies. As this vibration occurs, the needle vibrates a magnet between coils that induce the fluctuation of the electrical frequency, transmitting it to a loudspeaker, which in turn has a part known as diaphragm or cone that vibrates with the sound output. The diaphragm is a surface that moves in and out, creating variations in the atmospheric pressure and generating the sound according to the frequencies, thus perfectly reproducing the content of the sound image.

Theft of vocalized thought and visual thinking

There are several different ways and theories on how it's possible to capture audio that travels through the cortex via electromagnetic re-radiation. However, the way the vinyl record works is easy to understand. Consequently, the theft of vocalized thoughts and visual thinking occurs in a similar way to the process that we saw above. Nonetheless, the sound design isn't sculpted on a physical medium such as vinyl — it's being dynamically *drawn* on the return wave that is being re-radiated from the target's brain. Then, the sound design is formed (modulated) in this carrier, or commonly known as the return wave.

To summarize: the input signal leaves the transmitting antenna in a certain "format". When faced with the target's brain, this signal is reflected — after gaining processes due to constructive interference — and the return signal receives small changes in its shape. These changes occur when interacting with the Medial Geniculate Nucleus (MGN), where signals of all sounds processed by the brain travel — including the vocalized thought. Thus, the return signal is received with these changes, which are nothing more than the representation, the "sound design", of vocalized thoughts.

Using several complex algorithms and mathematical calculations, such as the Fourier transform, the computer system is able to infer the data with certainty, decode the received signal in audio and transcribe the audio into text in a matter of milliseconds. In this way, absolutely everything that the target is thinking in a vocalized way is extracted from their brain. The return signal has the *sculpted sound design*. The difference between the signal input and the return output makes the software that governs EMR calculate these frequency differences and digitally create the sound.

So, still using this analogy, the systems that receive and decode the return signal would be equivalent to the operation of the record player. By tracing the grooves of a record, it's able to electrically create the identical signal (the exact frequencies of the sound). However, instead of reading the sound of an "acetate", the format of the sound is modulated with a return carrier containing the frequencies — the vocalized thoughts of the target.

And to make matters even worse, it can also capture all audio from the environment in which the target is inserted. Everything that is processed by their hearing is also captured and filtered in a separate audio data stream. Incredible, isn't it? It's the purest scientific art in its utmost expression of human rights violations.

Keep in mind that vocalized thought is nothing more than your thought expressed in words. As soon as the words are spoken in your mind, they travel through the same brain *ducts* as your hearing does. That is, as soon as you hear your intimate soliloquy — the vocalization of the interpretation of reality — and express it in words, such words are heard by your "self". That way, everything the target thinks and hears is stolen every millisecond.

CHAPTER 2.2

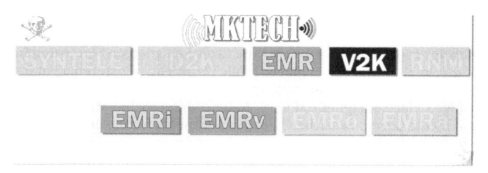

V2K – "VOICE TO SKULL", INTRACRANIAL VOICE, MICROWAVE VOICE OR MICROWAVE HEARING EFFECT

"The horror capable of driving a human insane in a short time."

— Dr Robert Otto Becker [12]

In the previous chapter, we've seen how to extract an individual's thought straight from the mind as the thought is processed by the responsible cortical areas. So now we're going to look at how to insert sounds and voices directly into the brain. This is the second sub technology used in the Mind Control Technology (MKTECH) system.

Although this technology is being discovered by the general public just now and in the form of psychotronic weapons, V2K is an old technology. It comes from the mid-60s and 70s, but it started in the creation of the radar. The first radar was built in 1904 by Christian Hülsmeyer in Germany. At that time there was no practical use for the device of low accuracy and inefficient echo detection system that was difficult to build. In the early stages of radar development, there

---

[12] - American neurophysiologist, author of several books related to the subject, such as "Electromagnetism and Life", responsible for several researches in the field of electrophysiology and electromedicine.

were no devices capable of producing radio waves at the desired frequencies and with reasonable powers. The common vacuum tubes weren't yet sufficiently developed. Even if they were, they still faced serious difficulties in producing waves at very high frequencies.

The first practical device capable of producing high frequency signals for radar applications was the Cavity magnetron [13], created in 1921. In 1934, French engineer Emile Girardeau registered a patent for his work on a multi-frequency radar system. As Pierre David reviewed electromagnetic theory, he encountered the study carried out by the German and then began experiments for the development of a high frequency radio wave detection system, efficient for the location of airplanes. Simultaneously, Henri Gutton and Maurice Ponte managed to create a detection device that worked with great precision.

Still in 1934, in the Soviet Union, engineer Pavel Kondratyevich Oshchepkov, who received the Order of Lenin, Order of the October Revolution, invented RAPID, a radar system that could detect the presence of an airplane 1,865 miles away. In 1935, the United States attained its first single-frequency radar through the work of Dr. Robert Page. The first radio telemetry system was installed on the Normandie ship in order to locate and prevent the approach of obstacles.

At the beginning of World War II (1939), Watson-Watt improved and developed new technologies by using the fixed and rotating telemetry system. There was a more rapid evolution in radar technology during the Second World War. Both the British and the Germans were involved in a race to produce larger and more sophisticated radars. The race was largely won by the British who knew how to use more effectively the radar system they had.

But what is a radar?

Radar means: *Radio Detection and Ranging*. It's an electromagnetic sensor that detects and locates targets by means of reflection,

---

[13] - The cavity magnetron is a vacuum tube present in microwave ovens and is responsible for transforming electrical energy into electromagnetic waves.

composed of a transmitting/receiving antenna of super high frequency signals (SHF). The transmission is an electromagnetic pulse of high power, short period and very narrow beam. During propagation through space, the beam expands in a cone shape until it reaches the target being monitored. The beam is then reflected and returned to the antenna as it creates the image, the echo. There must be no obstacle between it and the target for the radar to work. The wavelength must be as short as possible, within certain limits, which implies very high frequencies so that we can detect small objects.

Thus, the greater the power, the more echo we have and the easier it will be to detect an object, however far and smaller it is. It's known that the distance from the object can be easily calculated by the constant speed of pulse propagation, and by the time of arrival of the echo. It's also possible to know if the target is moving away, or approaching the station, due to the Doppler Effect. That is, due to the frequency gap between the emitted signal and the received one.

2.2.1 - The history of the Intracranial Voice

The microwave hearing effect, or intracranial voice, is a well-established phenomenon that began in the 1960s through research by Dr. Allan H. Frey — an American neuroscientist who studied the phenomenon and was the first to publish an article on the topic — involving radar hearing. It was found that roughly three-tenths of a watt per square centimeter of skull surface is required to generate the clicks from which the voice is synthesized. This is considered a primitive form of digital audio. Other notable contributions to the early development of the microwave hearing effect include Dr. Joseph Sharp — Ph.D. in Psychology and Neuroanatomy from the University of Utah.

The voice to skull transmission was successfully demonstrated by Joseph Sharp for the first time in 1974. It was performed with Dr. James C. Lin's pulsed microwave transmitter at the University of Utah at a seminar presented to the Engineering and Psychology faculties. The transmitter was then described in the *American Psychologist journal* in 1975. The experiment is based on the fact that a medium-to-high frequency microwave radar pulse can produce an audible click in the auditory cortex of a person who is *resonating* with the pulsed signal. This effect has been called "radar hearing" since World War II when it was first detected. In 1976, the *Los Angeles Herald Examiner reported* that the Soviets were also conducting extensive research on the microwave auditory effect. This issue was brought to the attention of the U.S. Defense and Intelligence Agency and it was described as "words that appeared to be originated from inside a person's skull".

Dr. Joseph Sharp used a computer to produce a microwave radar pulse transmitted every time the speaker's voice waveform oscillated, every time the sine wave crossed the zero reference in the negative direction. The voice was modulated into a new signal — into a microwave pulse capable of being demodulated by the brain. This experiment is neither secret nor classified; it's available to the general public. On the other hand, the continuity of the experiment by the military, as an electromagnetic or psychotronic weapon, is in fact classified as Top Secret and is unavailable for public consultation these days.

The result was that when Dr. Joseph Sharp sat near the device in front of the microwave transmitter emitting pulses — as shown above — he was able to hear a "robotic" voice counting the numerals from 0 to 9. Afterwards, the experiment was discontinued by Dr. Sharp and became a state project. It started to be developed by military and intelligence departments such as the CIA and KGB. The race for silent, invisible weapons that affect all living organisms on the planet with a focus on the bioelectricity of the most developed brains (mammals) had begun. Most researchers from the mid-1960s into the early 1990s have concluded that bone conduction is the physical mechanism that allows for microwave hearing. Bone conduction is just a secondary pathway for sound transmission in humans and most other mammals.

AUDITORY SYSTEM RESPONSE TO RADIO FREQUENCY ENERGY—FREY

With this arrangement, however, the subjects invariably set the filter to cut out all frequencies below about 5KC audio and wanted maximum bandwidth to the high end.

Deaf Subjects: Only transmitter A was used with this series.

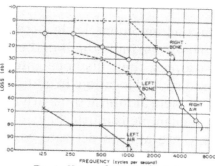

Fig. 1. Hearing loss in Subject 1.

Subject 1. The right ear was moderately scarred and thickened (Fig. 1). The left ear showed a clean radical mastoidectomy cavity. Subject 1 did not hear the RF sound even when the power density was 30 times that needed for the normal threshold.

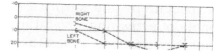

venous neomycin. A tinnitus persisted and w; described as sounding like the hiss of escaping gas. The subject (Fig. 3) did not hear the RF soun

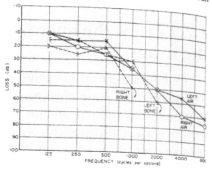

Fig. 3. Hearing loss in Subject 3.

Subject 4. This subject was not a clinical case a1 had normal hearing. He accompanied the investigat as an observer and participant in the experiment. l reported that he could not hear the RF sound. , audiometer check revealed the results (Fig. 4).

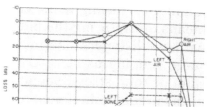

Figure 2.6 *Excerpts from the study by Allan H. Frey (1960) describing the microwave hearing effect, also known as Frey Effect. Research showed that a 1.3 GHz radio frequency transmission with peak power of 267 mW/cm2 could induce auditory sensations.*

The microwave auditory effect is a scientific fact. Existing radar units can be modified to transmit a beam of pulsed microwave energy into anyone's skull at ages ranging from 0.8 to 100 years, and across the planet, which causes sounds of any nature to be interpreted by the auditory cortex. The sounds, which appear to originate from inside, from above, or from behind the head, are transmitted to the inner ear through bone conduction.

There is no external noise. Other people around the person who is configured to receive the pulse cannot hear the microwave content, the sound itself, and the pulse can be applied thousands of miles away from the target.

2.2.2 - How the microwave voice affects hearing and the brain

The auditory sensory system plays a fundamental role — regarding phonation — in communication between individuals of the same species. It's used to locate sound sources, their distance, intensity and position, temporal frequency and spatial location. In addition to this main factor, other more complex neural interactions are intensified through sound stimuli. A good example is music. Few musical notes combined in a harmonic way are able to activate complex mechanisms in human beings, from evoking memories to intensifying feelings. Music manages to modify the person's physiological representation, which can lead them to a state of sadness or profound joy, causing tears, chills or a rapid heartbeat. It serves as a means of capturing the author's feelings and communicating them with the listener. Like music, sound stimuli can alter our mental state, keeping us alert, creating expectations and surprises, deep calm or total stress. Sound is capable of activating several mechanisms in the brain linked to all kinds of sensations. When this communication is made through verbal code by using language and voice — sound as a transmission channel—the social communication mechanism, which includes processing meanings, comes into play.

The word conveys information, the process of interpreting the coding of the word recruits almost all areas of the brain. Verbal communication for humans is inserted in a larger context than in other animals. The auditory cortex is surrounded by several Brodmann's and Wernicke's areas, which are essential for the interpretation of abstract aspects of speech, as well as the emotional content found in the contralateral hemisphere and neuronal patterns of speech that are located in Broca's area.

There are at least three pre-receptor phenomena of great importance for auditory mechanotransduction:

1 → first, the length of the external auditory canal and the mechanical properties of the middle ear determine the range of temporal frequencies transmitted to the mechanoreceptor cells located in the inner ear.

2 → second, the amplification of the acoustic signal occurs in the middle ear. It reaches the cochlear fluids where the mechano-electrical transducer currents in hair cells are located.

3 → the properties of the basilar membrane cause the acoustic pressure wave to propagate to the apex of the cochlea. High-frequency sounds localize near the base of the cochlea, while low-frequency sounds localize near the apex.

This frequency distribution is called cochlear tonotopy, in which each tone is represented in one place in the cochlea. It's a complex mechanism responsible for the recognition of frequencies, so that external sounds become audible. High, medium and low external stimuli are then converted into recognizable signals due to a mechanical property of the basilar membrane. Cochlear nuclei participate in the temporal coding of acoustic stimuli, which identify qualitative properties of sound.

The movement in the peak vibration is amplified up to 100 times by the effect of active filter provided by the electromechanical contraction in the external hair cells. From here the neural pathways are connected to various areas of the brain distributed in the bulb — located in the "thalamic bridge" and the cerebral cortex — and the bulb receives sound stimuli of certain frequencies. Frequency range of 20 to 20,000 Hz, with a peak of between 500 Hz and 5,000 Hz.

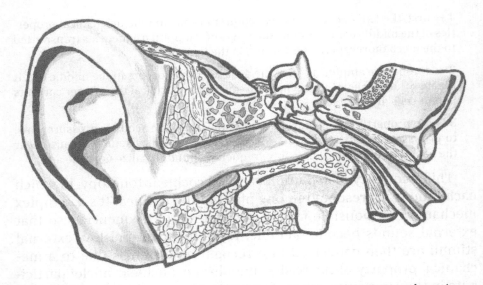

Figure 2.7 *From 16 Hz to 20 Hz to 20 kHz to 32 kHz - Minimum and maximum frequency ranges learned by the human ear.*

Human hearing can detect the frequency, intensity and timbre of the sound. The field of audibility, however, is quite limited both in terms of frequency and intensity. Very low intensity sounds are not perceived. However, the most intense ones cause a sensation of pain, so the V2K — Intracranial Voice/Microwave Voice — is capable of emulating any type of auditory sensation felt by the human being within these frequency ranges.

The primary purpose of this technology is to remotely send data and sound information over long distances directly to a person's brain without others around them being able to hear what is being transmitted. It's a silent conversation, because the target hears what is being sent, even if the environment is very noisy. To understand the level of intelligibility with which the sound reaches its destination — the target's mind — it's even possible to hear the microwave voice if the receiving target is at a large musical event with very high decibel levels. In order to illustrate the situation based on real events that occurred with some targets, let's take as an example a heavy rock concert or any other music genre with high levels of decibels.

The Targeted Individual goes to the event, enters the venue and finds a good spot amongst the crowd. This place may be close to the speakers, on the sides or in front of the stage — it doesn't really matter. Even with the guitar, drums and public excitement, the target will still be able to hear the penetrating messages sent by microwave pulses. The music captured by auditory systems cannot drown out the sound sent via V2K.

Another example is a crowded stadium like Maracanã, during a championship final. Despite the singing, public excitement, shouting, bass drums and fireworks, the target will still be able to hear everything sent by V2K, because of its invasive sound transmissions of good fidelity. Microwave sounds overlap all sound stimuli processed by normal auditory pathways resulting from an unusual factor that may cause confusion at that moment. The V2K stands out from the natural sounds that reach the brain and can be used as a charging source, as it uses the energy of the noises captured by the individual's ear as an impulse to become even more intense. This phenomenon is going to be discussed later in detail in chapter 2.7.

Another example of how the brain gives processing priority to microwave voices over other sound stimuli: during a street Carnival, as one walks among the crowd, amid bands, classic Carnival songs, drums and thousands of people going up and down the slopes with the Carnival blocks. It's impossible to talk to the person next to you due to the volume of ambient music. I believe that most people have had a similar experience. In order to communicate successfully, you have to get very close and practically shout in the ear of the person who is going to receive the message with enough intensity to overcome the noise of the environment. In this case, the microwave voice stands out, and can be heard clearly even in the midst of a noisy street carnival.

In essence, V2K is an extremely efficient [14] and bizarre form of communication. It has several properties and functionalities if applied correctly: in silent communications for soldiers in the field, between astronauts and Earth, or orders for agents in enemy territory, as well as silent communications between two people, for example.

---

[14] - Deaf people aren't able to hear the microwave voice, as it depends on the maintenance of the auditory system to be captured.

However, this technology is most commonly used as a form of debilitating torture, being one of the most lethal [15] psychotronic weapons, capable of continually sending messages to the victim with no possibility of being turned off or blocked [16]. The mere fact of staying alive becomes unbearable. Voices, screams, howls, cell phone ringtones, background music, alarm clock tones, electronic and metal sounds, constant water noise or any other sound that the operator [17] behind the technology wants to convey causes delirium, spatial confusion and emotional lightheadedness. With these unwanted noises invading the individual's brain 24 hours a day, anywhere in the world, the torture becomes extremely serious with direct consequences in all aspects of life. It makes the brain work beyond capacity, usually creating a state of vigilance, negative expectation and permanent alertness. It redirects thoughts at inappropriate times, drawing attention away from the reality around them.

V2K also has the ability to evoke memories in an improper manner. Words are one of the ways to access memory. In addition, it keeps thoughts chaotic when the body needs to rest, causing sleep to be less fulfilling or even interrupted and, consequently, it leads to physical and psychological exhaustion.

One of the serious problems of using this technology lies in the fact that it exploits the brain's inability to distinguish a legitimate sound coming through normal channels from a sound sent via microwaves. Thus, this technology is used indiscriminately, especially for torture. The attack is invisible to the human eye, leaves no trace and cross most urban obstacles. Moreover, it's not possible to capture the origin of the signal if it is disabled at the emitting source after the signal is processed by the target's cortex. Expensive, inaccessible equipment is required for most people to block and track

---

[15] - Although directed-energy, psychotronic and electromagnetic weapons are classified as Non-Lethal Weapons (NLW), they're extremely lethal and dangerous. Saying that they aren't lethal goes against this controversial concept and doesn't reflect reality as a whole.

[16] - One of the few safe places that would make the attack unfeasible would be a room completely shielded against electromagnetic waves. An extremely expensive room that isn't easy to find and few people have access to it.

[17] - Operators are people who use this technology and are part of a network involved in several serious crimes around the world. For the moment we're going to use the term "operators" to name all the parties involved. As we move forward, we're going to specify more precisely the role of each participant.

this signal. The victim will be at the mercy of the technology operators and, consequently, of the content and its frequency, the intensity of the sounds and the messages they want to transmit. Hence, it has become the perfect instrument to inflict physical and psychological pain on individuals anywhere on the planet. Criminals don't even need to expose themselves. So, operators can settle in a remote base and use a transmitting antenna to send the V2K content, taking advantage of the infrastructure capable of triangulating between satellites, antenna and target. There are unscrupulous people who make a living from this type of attack. We're going to learn about the operators behind the technology in chapter 5 in the next volume.

Insults, bad words, sexual and moral harassment are the most common messages sent to the victims of this technology unknown to the majority of the general public. These messages are always addressed in a similar way by the operators who send such messages from a remote position and in complete anonymity. Torture using V2K is so intense and stressful that it can make a target collapse in a matter of days, leading to suicide, to the accomplishment of some serious crime of interest to operators in exchange for the end of the torture, or even the possibility of being manipulated under intense torture with mind control techniques that could make the victim a possible remote killer [18].

Depending on the objective of the operators and the intellectual, socioeconomic and religious level, in addition to general knowledge and other factors that make up an individual's personality, the microwave voice can be used in several ways, from causing intense psychological and physical pain, performing information theft, simulating the voice of a mythical creature or even making the victim think they are getting symptoms of a mental disorder called schizophrenia. After all, V2K is able to simulate it perfectly.

The same symptoms of the disorder are also observed in targeted individuals subjected to long-term torture. Nevertheless, the external symptoms and the internal cause are artificially provoked by

---

[18] - Remote killer is one of the terms coined to describe people who commit murder under the direct influence of this weapon. The act may occur due to systematic manipulation in interpreting reality in the target's mind based on brainwashing techniques under intense torture. We're going to see in more detail the winding paths to reach such a state.

third parties via electromagnetic waves. In the course of the attacks, some victims may feel the need to consult a specialist in mental disorders. Unfortunately, they're likely to receive an inaccurate diagnosis, and the symptoms will be wrongly assessed, as it happens with most targets today due to several factors. The most important of them is the complete ignorance of this technology by those responsible for diagnosing these diseases. They diagnose the patient's clinical condition, which are associated with the reported symptoms, in addition to prescribing from strong medicines to unnecessary hospitalization.

Some symptoms of V2K victims are artificially created by repeated, systematic and constant microwave stress. A more in-depth analysis of the subject is found on Chapter 2.3, in which I coined a new term: electronic schizophrenia. In this chapter we're going to focus on details that produce pseudo-hallucinatory symptoms, especially auditory ones, persecutory delusion, among others.

The main purpose of using intracranial voice is to cause physical and psychological pain to the victim, to prevent them from reasoning as they did before they became a target. Stimulation of the brain makes it work under constant stress, which makes the victim develop mental health disorders. In addition, it makes impossible for the individual to perform daily tasks, mainly of an intellectual nature that requires calculated reasoning, constant attention or high cognitive level. How to deeply think about a certain topic, how to reflect abstractly, how to study some subject or perform the job itself are some examples. The main objective is to neutralize someone, leaving them unable to react to attacks and simultaneously making a complete degradation of their quality of life.

This technology also serves to extract information from the target's minds, activating memories with auditory stimuli and making the brain think about subjects that are of interest to operators. It directs the victim's thoughts with the auditory stimulus, activating the memory regarding the desired topic, and can be picked up by the Electronic Mind Reading (EMR) as the thought is accessed by the target inside their own mind. These advantages occur mainly at the beginning of the attacks, in which the surprise factor is used, and they exploit the total lack of knowledge about the existence and functioning of this technology.

Victims of this weapon may experience unique auditory sensations. Such sensations will be interpreted by the brain as if they come from different sources of different intensities and directions. It simulates well-known effects from movies and video games; however, they are generated directly in the target's brain. The scientists who turned the V2K into a weapon of psychotronic torture had the idea of incorporating prominent surround effects, and they mapped the areas and neural network of the entire sound path in terms of intensity, position and sharpness. How the brain interprets the sense of position and distance, angle and degree of sound was then tested. So, they used this information and performed a kind of reverse engineering. They verified the areas activated by certain types of sounds and how these sounds are interpreted by the complex auditory cortex. Then, they started to electromagnetically stimulate the same areas of the brain in order to deceive it and, thus, to create a sense of virtual involvement.

I'm going to exemplify the way the brain interprets sound waves and how they are virtually simulated and incorporated into the reaction to V2K, thus deceiving the superior olivary complex — area responsible for the horizontal mapping of the location of the sound source in the plane. In the case of sound waves (mechanical waves), the difference in amplitude and time makes us sure of the position of the sound in space: right side, left side, center, diagonal, back. Now imagine a noise on your left side. The sound travels through the air and reaches your ear. Soon after, it arrives at your right ear milliseconds apart, but of a lesser intensity, since the sound can be refracted, reflected and dissipated by obstacles in the way. The sum of these different peculiarities of amplitude and time is what makes us feel the direction of the sound. Frequencies of responses regarding sonic signature features, delay in the sound-arrival time, among other factors, generate the sound interpretation that the brain presents to us after being processed. These same sensations are emulated by the microwave voice, activating the same areas of the brain equivalent to those activated when the legitimate sound (mechanical wave) of the same effect is generated and interpreted by the auditory cortex.

Another effect used by the entertainment industry, which is artificially induced by V2K, is the directional and positional effect of the sound. It gives a false impression of the location. For example, a delay of 250 milliseconds gives us the impression that the audio source is 30 degrees to the right or left of the center. So, the longer the delay time, the greater the sensation that the sound is being playing next to us. The shorter the delay time, the more centralized the sound will be.

Just as the interauricular time interval: if it comes from the right, it will reach the right first, then the left; if it comes directly from the front there is no delay; from the left, it will first reach the left, then the right. This delay allows us to locate the sound sources in the horizontal plane. Continuous and very loud sounds create sound shadows, allowing neurons specialized in intensity to locate the source of the wave.

* 20 to 20,000 Hz → interaural time difference.
* 2,000 to 20,000 Hz → interaural level difference — duplex theory of sound localization. Acoustic shadow.

A good horizontal location requires the comparison of the two sounds that reach both ears, but this isn't necessary for the vertical location. The vertical location of the sound source is based on the reflexes of the pavilion of the ear. This creates a completely positional, 3D immersive sound, directly into the brain, leaving the target completely disoriented and frightened.

Basically, these are some processes that create sound localization in the human brain. That way, the known positional effects have already been mapped and studied in the laboratory, and it was found the neural networks that are activated in the process, the areas of the cortex involved, the intensity and time in which these networks are activated and the way to induce responsible neurons sensitive to the difference in pressure and frequency. The neural configuration resulting from the process is emulated via V2K. It makes the target believe that the stimuli resulting from the microwave come from different positions and intensities as a mechanical sound, thus deceiving the brain in all occasions. Keep in mind that there's no difference in the way the microwave voice and the mechanical sound are interpreted by the brain. That's why this feeds the confusion and increases the madness.

A very common sound sent time and again to the victim's head is the simulation of a person yelling and screaming at the top of their lungs from a window of a building: it gives the impression of being far away and echoes through the block. It may also be a smooth female voice with no sound effects, which gives the sensation that the voice came from above. Whisper-like effects that appear to come from behind the head in 180 degrees are also common. Some effects are able to imitate the inner voice, the vocalized thought of the target. The sensation is that the voice is generated right in the core of the brain, however, without having been mentalized by the target. This intrusive voice that comes from within the mind has a high capacity to access memories and create negative mental effects, as well as to direct the flow of thoughts, preventing proper attention and sleep.

Taking advantage of the evolution and popularization of home studios, technology operators can pre-edit or create filters for the channels on which certain voices will be sent. The effects used are electronically generated, but they give the feeling of being created according to the surrounding environment. It's worth bearing in mind that these are microwave pulses and radio waves, not real sound waves, and they don't obey the particularities of the laws that govern mechanical waves.

Although it depends on the level of knowledge and creativity on the part of the operators, in which practically everything can be done, the most common effects are:

- **Distortion:**

Modification of the sound wave form of a complex sound (timbre) due to uneven changes in the amplitudes of the components, leaving the voices lower or higher. It's widely used as the voice of monsters or heavy metal singers.

- **Resonance:**

A body can vibrate when it receives elastic vibrations from its surroundings, so that any structure in a building, like a wall, can oscillate.

- **Echo:**

It's the phenomenon by which the reflected sound causes another auditory sensation. Repetition of the original signal happens when the reflected auditory sensation occurs in a time interval longer to 1/5 of a second.

- **Reverb:**

It's different from the echo effect. While the reverb characterizes a continuity of the sound in the environment, the echo is characterized by its distinct repetition. Sound of 60 dB (moderate).

- **Effects in general:**

Cathedral, Hall, Arena, Degradation and Old telephone are also used for voice effects.

The operator decides how the target's brain will interpret the sound — the content that will be sent within the microwave pulse signal. The program that configures these parameters works as described below.

The first channel of a given voice will be configured so that the brain interprets it as a sound coming from the back of the receiver. Therefore, the target has the sensation that someone is on their back giving high-pitched screams as if they are coming from the apartments in the neighborhood. Second channel, front position; third channel, side position and so on. Each channel can come with built-in effects — depth, distance, echo, cathedral — all the known effects of the music creation and production studios. Technology operators generally begin their attacks on an ongoing basis using these emulated properties of the brain's ability to interpret microwave voice with these characteristics. **They adjust the system in order to create an auditory illusion in the target's brain identical to having neighbors screaming in the neighboring apartments.** It makes the victim paranoid with those who have nothing to do with the attacks, as the incessant fighting and demeaning shouts arise from the same position 24 hours a day — always coming from the apartments next to the victim's room.

Another widely used effect is to simulate several unknown people roaring in the buildings and streets around the victim's location. It gives the impression of a mob storming the building or a gang rioting and about to invade, usually at the crack of dawn in such a high, powerful and terrifying way that if it weren't an effect generated by the microwave voice and demodulated only in the target's brain, it'd attract the attention of the whole community. This isn't in fact the case in recurring situations like this worldwide.

The scream attack is used regularly, as it's known that noticeable fluctuations in the volume of the human voice, at a given intensity and in the frequencies **between 40 and 130 Hz**, automatically provoke instinctive reactions when processed. These reactions electrically conduct the brain to synchronized patterns in areas linked to pain and repulsion (insula, hippocampus and amygdala), making it virtually impossible to ignore them.

A great confusion occurs in the victim's head. It's practically impossible to distinguish between real screams and those generated via Intracranial Voice/Microwave Voice (V2K). In fact, those usual sounds of a city — people talking or playing football on a distant court, children playing, cars passing by, horns, motorcycles and animals, for instance — can merge with the microwave sound, driving the target crazy in a matter of days. This type of attack in which the targeted individual hears voices and screams in their own home, all day and all night, is part of a complex torture protocol with well-defined purposes and objectives, as we're going to discuss in detail in chapter 2.7.

Program interface

The program interface that configures the microwave voice has a field to choose which operator will be associated with a channel. In this channel, the horizontal/vertical position will be selected, along with the angle at which this voice will be interpreted by the target's brain. An effect from a list full of them will also be selected, as well as the sharpness. And this without taking into account the natural sound properties of the person's location that will be added to the voice when demodulated by the brain. If the victim is walking through a garage, for example, all the sound that reaches their ears is naturally perceived with an echo. The microwave voice will also be interpreted as if it were a genuine sound wave, as it has the interpretive characteristic identical to the environment. In this way, the voices will be interpreted with the same echo as the normal sounds of the environment captured.

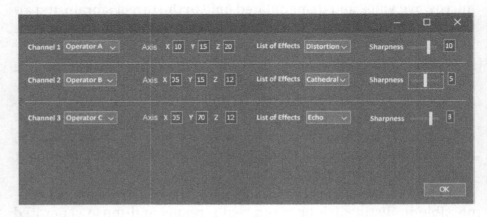

Figure 2.8 *Channel 1 - Operator A - Position <Levels | Horizontal | Vertical> + Effect [List of effects] + Levels of Sharpness + sound characteristics of the environment in which the target is located.*

Operators (or torturers) decide through these configurations how the target's brain will interpret the sound that will be sent within the microwave signal. They're equipped with a microphone of any kind and are connected to one of the several available channels, ready to rage. As soon as they utter the sounds, words and howls will pass through the system and acquire the properties previously configured and will be transmitted to the victim's mind thousands of miles away. Each operator will connect to a different channel; several voices will reach the target's mind simultaneously, each one of them assuming different characteristics and effects. Combining these two characteristics, position and effect, it's easy to induce the victim to think that there are neighbors yelling and screaming in the apartment above them or in the building next door, or even in the vicinity of their location, causing immense confusion and constant acute stress. It's impossible to distinguish the real ambient sounds from the artificial screams generated via V2K.

Generally, this technology can simulate ongoing, strange conversations that always seem to come from the direction of the victim's window, usually their bedroom or bathroom window. The electromagnetic wave that constantly strikes the target and mixes these channels sends this information to be interpreted by the target's brain in a format similar to 3D sound from movies and video games, but much more intense. It's practically impossible to discover the direction of the sound, since all directional interpretation is artificial and generated electronically.

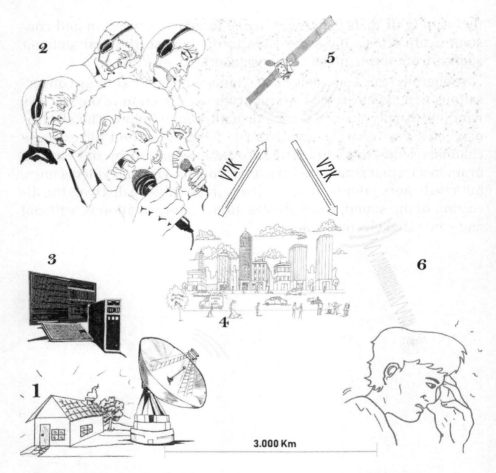

Figure 2.9 *Attack using V2K that reaches the target's brain wherever they go.*

1) In an isolated remote position, torturers/operators prepare to attack.

2) Each operator chooses a microphone and consequently a channel on which they shout and speak whatever they want to satisfy their histrionic desires and reach the target's mind using bad language.

3) The computer processes each voice channel according to the programming previously configured in the system.

4) An antenna sends the content transmission to a satellite, or to adjacent antennas, depending on the position of the target.

5) The satellite relays the attack to the target's position.

6) The target is hit by the blast — the microwave pulse—, causing enormous emotional and physical pain and confusion, which degrades their mental health. The

words that passed through the MKTECH channels are present in the content that was demodulated in the mind.

As they open a direct channel with the target's brain transmitting any type of auditory content in a continuous and systematic way, the audio sent by the operators may contain effects created in sound editing programs or something more rustic and older, however, equally effective. To mechanically create sound design without using electronic devices to generate sound effects, for example: a person using a megaphone or stadium horns, cupping their hands around their mouths and shouting, animal steps, boot steps, falling iron bars, saw or construction materials noises, repetitive music, and so on. The echo of the place where the sound is being produced is always captured and sent, which promotes more realism using the methodology of sound design of old movies — before the digital era — with improvised instruments and available materials. The sky is the limit for creating increasingly disturbing sounds.

Despite being a technology created in the mid-60s, it was only at the beginning of this century that the mass experiments involving mind hacks increased and are still in the execution phase regarding long-term consequences for the victim's brain and, as a result, their behavior through this torment. What is happening regarding the use of psychotronic weapons — like the V2K — is going to be clearer in the next chapters. Unfortunately, something that in the past was classified as science fiction today has become a sad and distressing reality.

Among the various techniques used to achieve and measure the interaction of the microwave voice in the target's brain, take this event as an example: imagine walking down the street normally — a common activity for most people. Suddenly, a repetitive instrumental background music starts playing, and is accompanied by a sensation that seems to be generated in the back of your head. With each step taken, this music is repeatedly triggered and synchronized with your walking.

This type of attack occurs constantly in the targets' lives, driving them insane. And to make matters worse, the harmful effects of this technology end up reflecting on the individual's behavior and social interaction. If they try to talk about it — to open up with close friends or family — they will immediately be considered mentally ill or advised to seek mental health treatment in 99% of the cases in which they mention strange voices or that particular song they're hearing everywhere. And all this regardless of the degree of intimacy and the level of relationship that the target has reached with the person. After all, this fact is automatically associated with mental disorders, as well as with the total lack of knowledge of the general public about this technology!

2.2.3 - How is the microwave voice demodulated by the brain?

The signals follow the set of frequencies that are exclusive to the target. Other people don't hear the microwave sounds intended for the victim. This reinforces the target's erratic behavior towards their relatives or close friends, as the signal is only demodulated when it encounters the numerical characteristics measured in their neural biometrics that was previously mapped before the attacks started. How the target's brain becomes unique is going to be covered later in the book, chapter 3.1. The audio resulting from the demodulation of the microwave signal "projects" audible sounds captured by their inner ear that are propagated by bone conduction and natural resonance of the neurocranium.

This is how the V2K evolved: from a microwave pulse [^~] [^~] [~^^~] that produced a small click in the brain to a profound violation of human rights, capable of performing unimaginable bad deeds in people's minds — something we've only seen in science fiction movies. The dominant and despicable characteristics of this weapon are a problem for the whole society, as they directly attack the creation of abstract and logical thinking, attention, memory and sleep. It completely distorts the cognition, mind and behavior of a person who becomes unintentionally and deliberately targeted by others.

Another serious problem is the level of implementation that has been carried out over fifty years, reaching a degree of sophistication. Today it has become one of the most impressive and devastating weapons within the MKTECH system and that's not the end of it! The current version of the microwave voice used today has several improvements over its predecessor with emphasis on a peculiar feature: it makes the microwave voice *blend in* with all the sounds around the target. It creates the impression of acquiring the sound characteristic of the environment, thus managing to deceive the brain and making it impossible for the individual to locate the radiation source. It produces the terrifying effect known as sound inside a sound, or voice within noise, voice within another sound. This is going to be discussed more deeply in the next chapters.

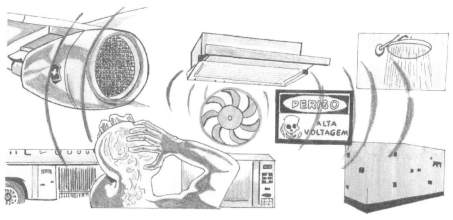

Figure 2.10 *Sound within noise.*

The technology used alone "just" sends sounds — voices, music, noises — of all types to the human auditory cortex and activates vocalized thought, mimicking the internal voice. It starts to confuse the Targeted Individual, because now there are several internal voices talking to the brain without the victim being able to block them or turn them off. Several voices of the mind — or vocalized thoughts — are simulating schizophrenia at the same time. Some victims end up developing real schizophrenic disorders, and this is one of the reasons for the silence regarding the existence of MKTECH in the last fifty years!

A number of reports concerning the use of this technology worldwide are coming to the fore from people who had their lives completely torn apart by this technology that destroys the human being's cognitive process. Now, think about the use of EMR combined with the microwave voice in perfect sync and without interruptions. Let's see what can be done with this combination. From then on, the technology starts to live up to its name: "Mind Control Technology".

Welcome to the definitive MKTECH technology: SYNTELE — Synthetic Electronic Telepathy, a factory of cognitive terror; the final frontier between the complete destruction of the human intellect and freedom, the mental privacy of the democratic society as we know it.

CHAPTER 2.3

SYNTELE - SYNTHETIC ELECTRONIC TELEPATHY

"There are several people listening to my thoughts and commenting on them inside my head! How is this possible? I can't think anymore, what's going on? I think I'm going crazy!"

— Anonymous Targeted Individual.

Artificial Electronic Telepathy, Synthetic Electronic Telepathy, or simply Electronic Telepathy, consists in the use of the EMRvi – Electronic Mind Reading (vocalized/images) to extract information — content of thoughts — and the V2K to insert data — voices and sounds — in order to activate neural processes connected to hearing. Then, a totally silent conversation between the technology operators and the target takes place. In other words, it's possible to send messages to the operator with the thought alone and receive a response via V2K without anyone in the vicinity being able to hear what is being communicated to the mind, nor the content of the thought being amplified and transmitted. It closes, therefore, a complete communication cycle.

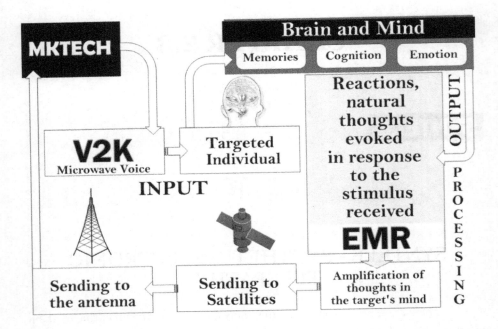

Figure 2.10 *Loop of Synthetic Electronic Telepathy:*

* Delivery of data by operators via V2K to the target's receptive functions → Input;
* Target reaction (thought generated in response to the stimulus received) and access to memory → Processing;
* Amplification (or reirradiation) and delivery of the resulting thought to the operators' base using EMR (Electronic Mind Reading) → Output;
* Processing of data received by operators → Reprocessing;
* Delivery of new content via V2K to the target → Input;

It's possible to send the content of thoughts to the other side of the world in a matter of seconds and get a response from the receiving operators and from the thought analyzer program according to what was sent. The essence of the technology consists in reading, in a non-invasive and remote way, a person's thought, amplifying that signal using sophisticated instruments, capturing this amplified information and sending it to receiving antennas at their destination. This destination is a place where the data received will be processed in computers and systems specialized in: a) transforming visual thoughts into images; b) transforming vocalized thoughts into text transcription (subtitles); and c) turning the sounds the target hears (including their own voice and vocalized thoughts) into audio. It'll then send a post-processing response from that set of received parameters directly to the same person's brain, using a microwave voice.

The Synthetic Electronic Telepathy (SYNTELE), if applied with the consent of the individual whose thoughts will be exposed, read, heard and seen by others, can be useful in various fields for peaceful purposes, as it is a way to help victims of cerebral palsy or with communication problems or those who have lost their vocal cords. It's interesting even for those with amyotrophic lateral sclerosis (ALS), a rare neurodegenerative neuromuscular disease that paralyzes the muscles of the body without, however, affecting brain functions. Stephen Hawking [19] would be one of the beneficiaries of this technology, since it would be possible to use vocalized thought to communicate in real time with everyone using his computer that emits a traditional robotic voice based on the reading of texts. It'd also allow accident victims to speak again.

Following this line of reasoning — that is, of picturing creative uses of this technology for a perfect, utopian world — imagine being able to talk to your partner without having to use your voice or mouth, without making a single sound, just using the vocalization of thoughts. You and your loved one at home, miles away from each other, amplifying the vocalization of thoughts and transmitting such thoughts directly to each other's brain via V2K. It'd be as if you were talking on a cell phone without the device itself. Better yet, no one around you would hear the dialogue or even know that you're talking to someone. It takes a lot of practice to learn how to control vocalized thoughts, and how not to send thoughts of momentary impulses or associative thoughts, which would be part of the cognitive processes in a normal conversation, but wouldn't be expressed in

---

[19] - Stephen William Hawking was a theoretical physicist, British cosmologist and one of the most acclaimed scientists of our time, responsible for several discoveries and theories. He calculated that black holes should thermally create or emit subatomic particles, known as Hawking Radiation. He also predicted the existence of mini-black holes. In addition, Hawking participated in the first developments of the cosmic inflation theory in the early 1980s. RIP † - 03/14/2018.

words. In the Synthetic Electronic Telepathy (SYNTELE) this is not possible. Everything that is thought is sent to the receiver.

It would be indeed an extremely useful and efficient tool for several impressive and peaceful purposes. However, it's only used for questionable goals such as torture, theft of information, espionage, illegal surveillance, fraud in civil service competitive examinations (Volume 2, chapter 9), terrorist attacks and murders. After all, it was conceived as a weapon of war, a "Non-Lethal Weapon", by MKULTRA's programs around the world.

The possibilities for using this technology are practically limitless. In addition to hearing and seeing the thoughts of the individual, it's possible to interact with them, making them know that their thoughts are involuntarily shared with third parties and causing immense discomfort, an affliction, an unprecedented sense of mental violation. We're not going to discuss the people behind the technology yet, nor their goals — that is for another chapter. Nonetheless, when MKTECH is used to neutralize, torture or murder, their participation is evident. The target's thoughts are forced to pass through numerous subjective filters created by the operators themselves who respond to every second of the victim's life. Thus, their thinking focuses on situations that have great emotional charges, including negative ones. This creates an unprecedented psychophysiological condition.

The target is subjected to the judgment and analysis of third parties for every thought, for every millisecond of their existence, for all the thoughts of a single day, for example: deductive thinking, inductive thinking, analytical thinking (consists of separating the whole into parts that are identified or categorized), systemic thinking (a complex view of multiple elements with their diverse interrelations), critical thinking, routine thoughts, automatic thoughts of memory maintenance and creative, philosophical, linguistic or abstract thoughts. In short, all thoughts — either voluntary or involuntary ones — will reach the operators' computers, completely destroying the target's reasoning capacity. It's extremely embarrassing and revolting, and it affects the person physically and psychologically.

Another creepy characteristic of the Synthetic Electronic Telepathy is its power to drive a person crazy in a short time. In a matter of days or hours, SYNTELE breaks down the mental resistance of most of the Targeted Individuals. The initial feeling of having your internal thoughts being collected and the difficulty in controlling what you think, exposing all of your cognitive intellect, shakes the strongest of human beings. That's why this is one of the main foundations of the Mind Control Technology.

Bear in mind that V2K (Intracranial Voice or Voice to Skull) alone has the potential to cause major damage to a person's mental and physical health. SYNTELE (Synthetic Electronic Telepathy), however, reaches another level in terms of torture and violation of the human being's cognitive privacy, because the voices are no longer just used to make random noises in the victim's auditory cortex in order to deprive the target's primary sense of natural attention. The voices interact within the mind, responding to the target's thoughts. So, the sadistic operators behind the analysis of these thoughts play, make fun, bully, disrespect and harass the strictly private cognitive processes of human beings, in addition to following some torture rules and protocols adapted for MKTECH, as we're going to see as we progress in this obscure technology.

Some authors, for the understanding of the reader who has never been in contact with MKTECH, like to make an analogy between Synthetic Electronic Telepathy and a "magic phone". That is, one can call anyone's mind anywhere in the world and basically say what one wants. The thoughts of the person who answered this "magic phone" would be the equivalent of the voice on the cell phone, but with an important difference: the phone cannot be disconnected, which keeps the caller in complete anonymity with unrestricted access to all functions of the receiver's "phone". It's a good analogy for us to begin to understand the power of this technology, except, of course, that magic phones don't exist.

Not only the violation of thoughts and the incessant noise in the victim's head — which is already a very serious transgression —, but the most worrying thing is the way in which Synthetic Electronic Telepathy uses its own communicative essence against the brain, the shortcuts used to perform, streamline and interpret the communication process, the way small cortical processing units interact with each other for this purpose.

The human brain is naturally configured and evolutionarily adapted to facilitate communication between peers. Therefore, the content of what is verbally transmitted affects the mind with a greater amplitude, reaching more cortical areas, since the influence of language that, although it's a product of the nervous system, activates, directs and stimulates the brain in a positive or negative way and is also the most effective way to activate the nervous system of others, facilitating communication even if it is via SYNTELE.

The words sent to the target's mind can mimic their internal voice, accessing memories or activating negative reactions, diverting attention in all daily activities, suppressing other external stimuli and forcing the target to maintain total focus on the subject at hand. It makes it impossible to inhibit intrusive thoughts, leading to a state of acute stress, usually for long periods. In addition, it forces the target to use all mental resources to fight the theft of thoughts and to defend themselves internally from invaders in a naturally instinctive way. SYNTELE employed as torture and theft of thought harms the formulation of thought itself, as it fixes the victim's thinking on a subject with a lot of emotional charge associated in a systematic way. It even damages the process of creating and fixing memory, since the constant flow information will be assimilated exclusively on the content of the attack and the impossibility of turning it off.

Operators can find a topic that causes deep sadness, resentment, regret, anger or any self-destructive emotion, and ruminate on this subject as much as necessary. They can keep the individual focused only on that content, which in a normal situation would be suppressed in the target's memory. Social and family problems, subsistence issues, that is, intrinsic problems of human coexistence produced by our social fabric are used by operators to incessantly attack targets. Imitating the target's natural internal voice, as well as activating it, creating access to thoughts and memories, exposing its contents completely, activating the mental circuits that process this information automatically and involuntarily, are periodic actions that are part of an efficient cycle of degrading attacks.

The brain constantly works with the internal voice for various purposes, including access to memories and their content. In the process of describing something, a situation from the past or an automatic response to a question, the vocalization of thought is used constantly to coordinate data and prepare the thought to be expressed aloud, following a line of reasoning. We need the vocalization of thought to talk, to request a subject and to use it in a conversation. When this process is violated and exposed, it causes an extraordinary negative impact for the individual. This internal voice isn't usually expressed in behavior; we just keep it silent in our cognitive inner self.

In the chapter in which I presented the EMR (Electronic Mind Reading), I listed several serious consequences for society and for the individual, using only the remote reading of thoughts in which both vocalized thoughts and images were sneakily stolen, without generating reactive behaviors from the individual being injured. Now with the complete two-way communication cycle, operators are able to stimulate the brain, listen to the target's reaction and continue to adapt the subject until the goal is achieved, be it torture, triggering memories, accessing hidden mental data, discovering secrets, generating negative physiological reactions or simply enjoying others' suffering. Along this path, we can point out other serious consequences for society as a whole.

So, the question remains: what would you do if you had these weapons in hand and could hear the thoughts of anyone on the planet? Would you contain your curiosity to violate the privacy of others? The operators behind the technology have no qualms about it. The power to communicate with a person's mind at a distance, silently — consensually or otherwise — opens up several possibilities that are unprecedented in history. I'm going to list some of them and I ask you to do a mental exercise yourself. See the infinite possibilities of using this technology for peaceful and beneficial purposes or for war, chaos and pain.

- **Used to defraud civil service competitive examinations:**

Delicate and complex subject that requires a separate chapter, which is contained in volume 2 of the book.

- **End of privacy:**

All the surprise and mysteries of a relationship, for example, will end when this technology becomes accessible to everyone, which paves the way for surveillance and obsession. From inside my own home, I'll "call" the brain of a woman or man of interest and I'll follow their whole routine. I'll be able to harass them sexually and morally, as some organized crime groups do.

- **Used by religious fanatics:**

Religious fanatics don't need much encouragement to make their hecatombs. Their main goal already lures vulnerable people capable of committing deliberate acts, like murder. With this technology, the religious guru will be able to speak in people's heads and deceive them by posing as an enlightened or powerful being capable of such a feat, escalating fanaticism and assurance in the words of the supposed guru with "supernatural powers".

- **Other functions:**

* To make people think they're talking to mystical or alien entities;
* Possibility of anticipating attacks;
* End of any intellectual activity that requires concentration;
* To check if someone is really guilty of a crime;
* End of cognitive liberty;
* To discover a person's deepest secrets;
* End of meditation, Mindfulness, Yoga;
* End of millionaire games like POKER;
* End of any activity that involves mental secrecy;
* End of any type of exams, tests or individual competitions;
* End of any activity that involves information stored in people's minds;
* End of any industrial secret and intellectual property;
* Accomplices talking freely with criminal gang leaders in prisons.

And so, we can go on for dozens and dozens of pages just to list the occurrences that will be completely changed in society if this weapon becomes popular and accessible.

The EMR (Electronic Mind Reading) and the V2K (Voice to Skull) devices were already discussed in the previous chapters. Now the Mind Control Technology begins to take shape. We can no longer disconnect only the theft of information from the torment of the targets. Torture is confused with theft of information in our biological *hard drive*. It's no longer possible to discern where one ends and the other begins.

SYNTELE for torture, experiments and information theft

Artificial Electronic Telepathy provides the perfect weapon for intense mental torture and information theft. It provides an extremely powerful way to exploit and harass people. It's capable of making other people's brain think about a particular subject — induced by the invasive internal voice —, and of collecting the processed response to that stimulus within the mind, generating acute stress during the progress of the attack.

Human diversity offers a wide range of personalities, mainly focused on emotions. Some people have strong emotions after few stimuli; others need strong stimuli to reach the same emotional level. And this emotional thermometer shows at what level of stress in the face of the systemic attack the target will withstand even a complete collapse. This data is measured and attached to a profile, with several other parameters captured through the most varied tests hidden in the attacks. Within that same context, how long it takes for the target to reach their limit and for how long they will endure suffering until they mentally succumb. Mind and body form a system that interacts and influences each other. It's not possible to make a change in one without the other being affected. Consequently, body and mind will collapse.

SYNTELE deeply explores the human ability for social interaction, cognitive processes based on reciprocal actions that take place between two or more individuals in which the action of one of them triggers the response of the other and promotes some reaction in others. This culminates in several different results expressed in physical movements, words spoken or in writing, producing what we know as interpersonal behavior. The brain is always rewarded for pleasurable acts and sensations so that it can play key roles in the interaction of social behavior, leading to establishing a stable and permanent bond. Operators are aware of this innate characteristic and exploit these processes to make the most of this interaction between the character that is mentally created on the target, captured from the details of the microwave voices, to associate with a real person — the operator behind the voice and the target itself.

In the process, many secondary thoughts emerge and they dictate the reason for such an association. Fears, theories, memories, past events, future concerns and problems with relationships are revealed. All cortical processes that work behind the scenes are captured and used later as ammunition for verbal attack by the operators. They can use our internal processes against ourselves, against the "self", the person, the "spirit", the individual that is the result of this electrical synthesis. Do you remember when you were a child and someone read a bedtime story to you, in which your imagination created a world based on the reader's words? It's exactly the same principle, but it occurs in a more persuasive and severe way. Stories and fables affect our perception with their literal meaning. We speak, hear, see and feel.

Synthetic Electronic Telepathy operators use warfare tactics, called Psy Warfare. These are psychological warfare tactics — this topic is discussed in detail in other chapters — from military agencies around the world. They're able to drive people crazy, because in addition to forcing the victim to hear voices and sounds wherever they are, these voices interact with their thoughts. Now, operators see and hear everything the individual thinks in a passive way and they can also actively interact with their thoughts. It forces the target to think about what they don't want to think about. It creates blockages in thinking, changes in the course of thought and its pace, and accentuates or causes severe intellectual problems.

MKTECH (Mind Control Technology) is capable of imposing intense torture, especially when the victim doesn't understand what is happening and an abnormal reactive behavior expected by usurpers is developed. The use of V2K leads people to wonder if they are schizophrenic. However, SYNTELE goes far beyond that: there are many voices of different people interacting with the victim's thoughts, responding, playing and reacting every second, talking to each other, creating an echo in their thoughts, mimicking vocalized thoughts that appear to be a very sophisticated type of schizophrenia. As this weapon becomes more popular, chaos will reign. House walls will no longer be the limits of a person's residence; privacy will be completely extinguished.

The moment SYNTELE (Synthetic Electronic Telepathy) is connected to an individual's brain, all of their vocalized thoughts are monitored and analyzed by unknown people in an arbitrary way. It becomes an ideal and silent way to neutralize any inconvenient people or any disaffection. It can be used at any time, including to end other people's lives, in silence, without leaving a trail and without being criminally accountable for it.

I particularly never thought that such technology existed or that it was at the level it is today until I "see it with my own eyes". The reports and documents claiming that military and intelligence agencies have kept secrets for years are not false. When they say that a certain project is classified as Top Secret, you can be sure it is! And its final form will only be shown when it is needed in real combat. As we see the abyss between military technologies and advanced agencies, we — mere mortals — try to adapt ourselves to the online world that is very important for our civilization. Meanwhile, society turns its full attention and concern to privacy on the internet, the day-to-day communication and social media without knowing that the privacy of real life is completely gone.

Online privacy is obviously important. After all, our vital data travels through these networks on a daily basis. But the excessive focus on online life serves, in part, as a distraction from the real threat: the stealing of thoughts directly from anyone's brain anywhere in the world and at any time by individuals who have access to the MKTECH infrastructure. There is no help to lessen suffering or to protect our cognitive data from theft by others. While we try to accommodate ourselves as a society on the Internet, military technology focus on a new Internet of thoughts, known as Brain Net, Deep Brain Web or Dark Mind Web.

The brain, or mind, if compared to a computer or the systems that make up a network, resembles an old and discontinued operating system — in terms of security of access to information — without any type of protection against attacks, easily hackable with infinite vulnerabilities and full of bypasses and backdoors [20]. And to make matters worse, we'll never have any kind of native defense against this weapon, unfortunately. It's a system in which improvements or updates that prevent security breaches will never be implemented. No *patch* is on the way.

So, when a person becomes a target, a simple walk on the street or any solitary activity in which they are left only with their thoughts becomes something painful, complicated, tiring and extremely stressful. Now the individual can no longer maintain the normal cognitive functioning of their mind as they always have. The simple act of contemplating nature, thinking about personal problems, making projections, watching people or the time go by, reading a book or watching a movie, become subordinate activities and practically cease to exist. Their thinking and focus are now on the issues that mind hackers are trying to bring to the fore, while attention is totally focused on confronting these invaders who become part of their life 24/7.

The strategy of attention deflection is purposeful, as it is known that attention enables information from the external world to enter. So, for stimuli to be processed, they need to be perceived, and everything we perceive depends entirely on our attention. Attention directs our focus, by inhibiting internal and external elements that distract us and by prioritizing what is most relevant at the moment. In this case, the brain must choose the content sent via SYNTELE. Then, the irrelevant, non-priority stimulus, which should be inhibited, becomes the main focus. On the other hand, the stimuli that should be given priority — work issues, social interaction, spouses, children, relatives and friends issues, attention in the surrounding environment for self-preservation, learning, reading, among others — become secondary. These stimuli aren't stored, computed, experienced; they're not perceived, supplanting important information,

---

[20] - Backdoor is a method used for securing remote access to an infected system or network. It exploits critical, undocumented flaws in installed programs and outdated software and firewall to open router ports. Some backdoors can be exploited by malicious websites through vulnerabilities existing in browsers in order to ensure full or partial access to the system by a cracker, for installing other malware or for stealing of data.

keeping strangers in complete control of their focus and attention, who will arbitrarily command what will be processed by the target: the information, the external stimuli so necessary for our life. This reverses the natural order of the cognitive mechanics of a healthy brain.

Imagine that you have to take a test, an exam or an interview that will decide your future and therefore requires extreme concentration and focus. You've been preparing for months on end; however, life happens and so do the most different types of situations, including unpleasant events such as difficulty at work, disagreements in a marriage, family problems, financial crisis and the loss of loved ones or the imminence of it. All of this can be a negative factor to be overcome at the moment. Natural sufferings of life must be abstracted during this period and the focus should be only on the future. But the moment these concerns unfold in your mind, they became internal thoughts and will be automatically picked up by the MKTECH equipment and operators will easily discover your concerns. Just think once — a mental reflex — and they will snatch all your afflictions. The moment you take the test, exam or interview of your life, voices of varying levels of clarity will begin to spring up inside your brain, in an unimaginable way! They'll raise issues of a personal nature, generating internal distractions that are impossible to block, with internal and external stimuli, reviving your concern, generating anguish and anxiety, completely changing the focus on what is important at the moment and shifting the concentration to a state of concern. They tear down your preparation and the test itself, and cause the primary feeling of anger, an overwhelming desire to shout: "Shut up!" for the harassment to stop. At this point, your test, exam or interview is practically compromised.

It was no coincidence that the technology was called "**The brain hijacker**" in Brazil. It lives up to its name, because the person whose mind is connected to the mind reading system cannot get rid of the invaders at any time of the day. They become a hostage to those operators who systematically interact with the victim, interfering in every day thought, whether banal or not, insisting on demonstrating that they're in constant control, asking for things such as "Take a certain action to free yourself". It often implicates the target in serious crimes (e.g., murders, robberies, sexual assaults and aggressions). However, the victim is never released from electronic captivity. The most worrying thing is that they start using the kidnapped mind as a source of entertainment, a perverse sport practiced with humans, known as **Torture for fun** — a torture-based fun. The target becomes a captive animal, the main attraction of a morbid zoo where visitors attack, flog and have fun with the howl of pain of this "human animal".

The seriousness of the use of this technology as a weapon and its devastating effects on the human mind make targets from different nations use a series of names resulting from the philosophical interpretation of the phenomenon, such as **Mind Rape**. In this case, the victim feels violated by several unknown people who do everything inside their mind. Operators disrespect their privacy, activate unwanted memories and have fun with the target's feelings and emotions. And the most striking thing is that all of this happens in all locations of the target's daily life, especially in their own home, which should be their private place and free from this type of harassment. Thus, one lives in a kind of **electromagnetic prison**, deprived of all senses.

Some victims also report that they're prisoners of their own thoughts, since those that should serve to refine perception and discriminate feelings are now used against the targets themselves, endangering human growth and its relative personal emphasis. Due to the size of the suffering imposed on the victim, some even compare this technology to Nazi concentration camps of World War II, in this case executed at a distance using electronic means — **Electronic Concentration Camp**. The sordid and servile characteristics are similar to those of the old camps. Other people and authors have also called it **The Silent Holocaust**. There are some other terms — I even created one. I called this technology **Electronic Halter** [Cabresto eletrônico in Portuguese]. Accordingly, the operators try to guide the target's life in a similar way to what they do to a mount. However, the strap is not fastened on the quadruped animal; it's fastened on the individual.

Try to envision being able to walk and go wherever you want to go, however, having all your cerebral cortical areas of primary external reception and internal processing — your core, your processor — invaded by hackers in a remote way. The target cannot walk through a park, or a pleasant place, nor contemplate the landscape and have that inspiring insight to deal with life, work, family or friends; it's no longer possible to renew their energies, to take an exam, to study, to work, to sleep and to wake up normally. Even the progress of society is jeopardized in the long run. If we stop thinking, if we're afraid to think or if we withdraw our thoughts at all times, we won't evolve.

To illustrate the danger we're all in today, imagine a genius like Albert Einstein. He used his imagination — his visual thoughts — to come up with theories, including the theory of relativity, which revolutionized the whole way we understand the laws that govern the universe, the gravity of light and space-time. If he had his mental processes, that is, his intellect and his creativity hampered by this weapon, hindering his internal travels through the seas of the imagination, he'd never have stood out in his days. We can mention other geniuses who might never have changed their era if similar technology existed, such as Tesla and the creation of the electrical system that provides light to our home; paintings by Michelangelo or Da Vinci that make our eyes and souls shine with excitement; and Newton's equations and laws. How many geniuses or people with potential may never be discovered because of this terrible weapon?

These inspiring emotions and sensations coming from an external stimulus will be blocked and replaced every second by SYNTELE interference, directing thoughts to something negative, and diverting the individual from free thinking. They experience a deep feeling of emptiness, as they lose the ability to think and, consequently, to enjoy the wisdom of life. This attack that has claimed countless lives across the globe is also called **invisible electronic prison.** The technology simply takes over the primary thoughts and reception systems of hearing, as we've seen in the previous chapter. It's worth mentioning that hearing, after the sense of sight, is the most powerful sense we have in terms of access to cognitive functions, such as memory, focus, concentration and sleep. It's directly responsible for interfering in the functioning of those functions. And the people behind the technology — those responsible for implementing attack techniques — are called thought-eating worms, neural electronic parasites, idea-sucking *amoebas*, sophisticated 21st century thieves or even **organized professional torturers**. In the following chapters, we're going to discuss more precise terms regarding the people behind this electronic apparatus.

At the height of mental degradation, given the efficiency of this torture, it can be decided whether the victim — within the spectrum of their reactions — is a likely candidate for suicide or a potential murderer. The decaying process occurs slowly and the target doesn't even realize that they're being led to one of these endings, which becomes tangible thanks to several tactics used to restrict basic aspects of life. It affects your well-being and health, as it hinders your cognitive ability — your thoughts — and steals any product or idea that may be created. That's why we called it **"Mind Control Technology"** with the proviso that until the present moment it's not possible to directly control a person in a waking state (awake), transforming them into an empty brain without a "soul", a "self"; and assuming command, just like flying a drone, guiding a video game character, controlling a puppet or similar to what happens in nature with the fungus *Ophiocordyceps unilateralis*, the zombie-ant fungus. This species of fungus is found in the Brazilian jungle and takes over an ant's mind [21] and body and directly controls its actions.

However, the torture techniques used in the MK-ULTRA experiments (volume 2, chapter 4) from the 50s to the 70s, with today's advanced technology, turn into a dance of death adapted to the 21st century. They're sufficient to make people commit acts of atrocity, such as self-harm, suicide and harm to others. Moreover, there are individuals who were admitted to psychiatric institutions after being victims of advanced mind control torture by the MKTECH system.

---

[21] - The fungus attacks the central nervous system to take charge of the body. The ant is then forced to walk towards a humid and cold region to serve as food and reproduction of the fungus.

When the target is able to perceive what is happening, they stop thinking normally as they used to before discovering that unauthorized people are listening to their thoughts. The surviving victim starts to think about what they will think. It seems paradoxical, but it happens as a natural adaptation to the assault on the mind. It goes against the innate functioning of the brain, which is used to taking shortcuts to form reasoning based on memories. Thus, the person affected by this crime creates adaptive changes in behavior, alters personal expressions resulting from negative experiences of SYNTELE. After some time engulfed in this torture, the person will no longer be the same. Their behavior and their way of thinking will be modified in all aspects!

The clarity of V2K (sounds) in the victim's brain and the constant remote reading of private thoughts create stress at intolerable levels, as the operators continue to attack interpersonal characteristics, causing unusual activity in the brain, fatigue, and emotions that strongly affect the target, usually primary and primitive ones such as anger, helplessness in the face of a situation, resentment, sadness and disgust. It leads to a mental collapse, which faces declines, growing fears and uncertainties over the days, exacerbating the primacy over mediocrity. The art of thinking for oneself is lost. They deliberately create conflicting ideas that cause confusion and disparate conclusions, reaching the goal of generating hasty actions on the part of the target.

The pleasure of wicked and sadistic torture directed at targets is noticed in the tone of their demodulated voices. Operators are pleased to violate the privacy of their victims, reading their minds and commenting on everything the targets think in an effort to demonstrate as much brutality as possible. They also make it clear that privacy no longer exists, which makes the target have the most diverse, decadent and pessimistic thoughts. And even those thoughts will be attached as causing great suffering. Thus, the feelings caused by these thoughts will be repeated until they don't reach the high initial effect anymore. After a while, however, these issues will return sporadically and will test the target's degree of adaptation to that stimulus.

We all know that going through all the setbacks of life alone is quite complicated. Try doing this with several people screaming inside your head, engaging in intimate moments of all kinds, not allowing you to fall asleep, interacting with all your thoughts, giving opinions on every decision, be it as simple as what to watch on TV, which book to read or about serious labor issues. It's insurmountable!

The technology used to torture the individual to death will initiate a maddening experience of being pursued by voices and sounds at any time of the day and in any location: at home, at work, on trips, at the gym, in meetings, without distinction of time priority, unless the victim manages to enter a fortified room, proof of electromagnetic waves, which blocks from the lowest to the highest microwave signals.

We now arrived at the most problematic and highest added value factor for operators among all aspects of the attacks: the target's inability to sleep when their brain is hijacked by the system. To be able to sleep — an extremely essential state for the proper functioning of the organism as a whole —, it's usually necessary to have a quiet environment where the brain is able to transition from waking to sleep. If SYNTELE is already highly harmful to the individual's physical and mental health during the day, the situation gets even worse at night. Operators keep the target's brain in constant turmoil. It's the equivalent of a party per night where several people are shouting, laughing, cursing and harassing "inside your head", interacting with thoughts, analyzing the target's memories and life, in addition to dwelling on concerns. They use tactics that cause prolonged stress, such as sleep deprivation. They also deceive the victim by simulating all types of auditory hallucination via the Intracranial Voice. This attack can be launched from very far away from where the victim is, since it's not a sound wave, although the result is similar to it.

Habits — reflexes of natural defenses against attacks, even unconscious ones — are created, in which the change in routine becomes evident. It ranges from subtle behavioral changes to total dysfunction. The victim may even develop a more serious post-traumatic stress disorder, which can consequently lead to alienation, fury, anger, guilt and suicidal thoughts.

The first behavior and change of habit takes place in the use of other sources of sound and visual attention. For example, spending every night sleeping with the TV on to distribute the attention of the words spoken in the mind, eliciting vocalized thought responses, agitating mental processes and keeping yourself alert, thus preventing the target from falling asleep. TV sound lessen these symptoms. In total silence, the victim's mind is completely dominated by SYNTELE (Synthetic Electronic Telepathy) and D2K (Synthetic Electronic Dream or Dream to Skull), which is going to be presented in the following chapters. In other words, sleeping in a complete environmental silence becomes impossible to occur. Sleep deprivation is a key factor for this torture to thrive. But we're not going to stick to torture techniques at the moment as this topic is going to be covered later on.

From the very first moment that this weapon is connected to a mind, the targeted individual's life takes on a new priority: regaining control of their mind and confronting the invaders. It usually happens in an intuitive way when the victim uses their vocalized thought to retort operators in an act similar to a shout, however, without saying the words out loud; a constant reaction after absorbing attacks during the day and night. Hence, a permanent soliloquy before an audience of unknown invaders is experienced.

Over time and depending on the frequency of the attacks and the intensity of the equipment, when operated at maximum capacity, the victim is able to *"feel the microwave radiation impacting the electrons and altering their moment, capturing the agitation over the entire brain, generating magnetic fields and energy charges different from those normally modulated by neurons. One has the impression that the entire cortex is 'energized', working at a voltage above normal"*. Poetic license aside, the situation is similar to putting your head inside a gigantic coil. You have the feeling that everything around you is shaking and that waves are crossing your body and playing with your entrails, muscles and nerves. This causes irreparable physical and emotional damage, as certain areas of the brain will incorporate the memories of this daily struggle during the emotional peak of this surreal phenomenon.

The memories and emotions related to this clash are so intense that the encoding of this sensory information and the stimulus is compared to a post-traumatic stress disorder. The thoughts concerning the event will fully occupy the neural working memory that in the future will become long-term memory. This makes your brain always access these more latent memories created electronically by intense torture caused by the Electronic Artificial Telepathy. This clash, in the long run, makes the victim develop a kind of electronically conceived catatonic schizophrenia, as we're going to discuss in the next chapter. Voices that come from everywhere penetrate each subsystem of the brain as they create and associate characters and a context to the narrative accompanying them. Thus, the imagination is automatically recruited based on their characteristics. In this permanent process, the victim's main cognitive functions are slowly modified. It's like a parasite that takes over every part of the cortex, forcing the target to live and to embark on the reality that MKTECH operators want to project as they take the victim along the tortuous paths of the script. **It's like a *cerebral* radio soap-opera**, as one tries to comprehend the situation and who is behind it.

The person going through this experience will never be the same. It doesn't matter if we try to reason that these attacks are "virtual" to the mind, as they don't actually hit someone physically in the sense of causing visible skin wounds or damage to organs and bones. Even in this light, it's far from being an epiphenomenon. The damage is visible and directly affects the entire health of the target.

As the neural combat drags on, the overall picture deteriorates, causing complete fatigue and headaches, which can lead to death from acute stress, or strokes in areas related to speech and hearing. Thus, the victims' negative feelings about this weapon are accentuated with each attack cycle. They feel hunted, persecuted, abused and humiliated by a person or people who refuse to give their names. These people contaminate the victim's mind at all times with the dirtiest and most perverse language imaginable. Added to all this is the repulsive act of making private cognitive processes public without the consent of the victim. There is an increase in the speed of mental activity that reflects on the body and the daily behavior of the target, who starts talking to himself or herself, gesturing and behaving erratically, a consequence of the violation of mental processes caused by MKTECH.

Bearing in mind that, in addition to carrying this imposing burden alone, the target has to continue living outside the catastrophic parallel universe created in their mind that is no longer an easy task. Life itself becomes quite complicated. An antagonism of extreme internal noise, but external silence for other people.

I – Battle tips for the Targeted Individual

Keep your position. Say that you will *not* give in, that you're aware that everything is generated electronically and that all tactics employed by the operators behind this psychotronic torture are already more than known. However, don't change your emotional/physiological state to anger, for example — which is very difficult to do in a situation like this, due to the nature of the constant violation. After all, maintaining your high level of permanent stress and shifting the focus on what you're doing to the topic they're projecting is the operators' primary goal. It makes you clash with them and feel anger, which completely degrades your life, while they laugh and have fun with your misery.

The accelerated growth in the number of victims of psychotronic weapon attacks in society is still well hidden and is confused with psychiatric illnesses and unsubstantiated beliefs, such as conversations with mythical entities. Most turn to drugs, give in to religion, commit suicide or atrocities and massacres, as they become "programmed" remote killers.

The surviving victims who manage to comprehend the situation and realize that some electronic device is interfering with the natural functioning of their brain — causing auditory hallucinations and reading the content that respond to thoughts — have no one to turn to. Unfortunately, there is no use looking for authorities, debating on the internet or seeking justice. No one will help! They'll feel like orphans and will avoid talking about it. They'll retain all surreal torture and cognitive violation for themselves, even if they're surrounded by people, because they will certainly be called crazy by friends and family if they share their experience with them. So, the victims end up carrying this burden alone, and they suffer the consequences of seeing their lives destroyed, especially the professional and emotional ones, in a purgatory of suffering and loneliness.

That is why it's so important to spread information about the Synthetic Electronic Telepathy and the psychotronic electromagnetic weapons. It's a complex issue involving several areas of science, which destroys our model of society as we know it for good. Hence the general public's initial reluctance to believe in its existence. As this is a weapon designed for war and kept secret with access restricted to a few, in addition to being improved for more than 60 years, it's vital for society that we have enough information to defend ourselves. That's indeed the purpose of this book: to inform people so that they're not caught off guard and become an easy prey for the torturers behind this technology!

Forget your electronic equipment, your cell phone, your camera attached to the TV, your cell phone or laptop camera. All of these will always be susceptible to invasion in some way. Compared to our human apparatuses, these are just superfluous devices. We're dealing with something much more worrying here: the access to human cognitive functions that occurs in a relatively simple way for those who have access to the well-established technology set since 1990.

Interaction on social media has become the new distraction for the population. It's a lasting alienation, mostly composed of much futility and unnecessary exposure, as we feed the system with personal information for others to enrich while the real threat surrounds us all. This is in fact one of the most complex technologies ever developed by mankind with the power to turn our entire society into dust.

2.3.1 - SYNTELE and Electronic Schizophrenia

"There's someone in my head but it's not me."

— Pink Floyd.

Erratic mental phenomena caused by the decay of the states of consciousness produced by the deterioration of primary systems, which is induced by the accumulation of electronic stimulus, the electronic catatonic schizophrenia, loneliness and the lack of attention in momentary issues are the more common collateral consequences in victims affected by SYNTELE for long periods.

I have no doubt that Artificial Electronic Telepathy can perfectly simulate schizophrenia — such as some of its characteristics and symptoms — in the minds of the targets. However, what is most striking is that technology operators use these symptoms as a kind of attack protocol. They try to direct the victim's behavior, targeting classic symptoms of basic, catatonic schizophrenia and some symptoms derived from other types of schizophrenia. That is, they create the symptoms artificially and electronically, thus generating the behavior of the target effectively schizophrenic, which can be confused in forensic analysis by competent professionals. It produces a correct diagnosis, but incorrect at the same time with regard to the cause.

Before we go any further, I want to make one thing very clear: at no point do I try to question the existence of these mental illnesses. They've been around since time immemorial — since the beginning of humanity, when electronics didn't exist. The overwhelming majority of diagnosed cases are of people with severe mental illness from organic causes not deliberately generated by external agents and in need of psychiatric and medical treatment. However, to make life even more difficult for doctors and psychiatrists there are people who are diagnosed with mental illnesses and people who are being attacked by neuroelectronic weapons, whose attacks generate practically identical effects to those mentioned above. Therefore, people diagnosed with severe mental illness must follow all medical recommendations and the treatment proposed by health professionals.

In this chapter, I intend to alert professionals in this area about this real threat that is never considered as cause of certain mental disorders. The matter is serious and must be taken into consideration in the medical field in order to diagnose this technology as a cause and the factors that lead to schizophrenic behaviors.

Operators use tactics related to schizophrenia, which create real symptoms of the disease in an infinite spiral for the patient or the target. Long-term attack strategies consist of highlighting the most common types of symptoms, such as several different voices with unique personalities within the victim's mind for long periods.

But what is Schizophrenia?

- Schizophrenia, schizotypal and delusional disorders (F20-F29):

Schizophrenia is a disturbance, a serious mental disorder that distorts thoughts and perceptions, loss of contact with reality due to severe disturbances in thinking and mood. The disorder manifests on a variety of levels.

Symptoms

When I saw the effects of the weapon in practice, it prompted me to investigate whether there was any disorder compatible with this phenomenon in the specialized medical literature; in other words, which mental pathology best fit with the symptoms observed. Certain symptoms reported in patients with schizophrenia caught my attention right away. In fact, these symptoms matched perfectly with the effects of SYNTELE on the brain, especially those that manifest immediately in the microwave hearing of multiple channels, such as listening to several voices and, subsequently, the same reactions manifested in the behavior caused by electronic torture.

For those who have already felt the influence of neural weapons in their mind and for those who are just reading about it, observe how some disorders resemble each other in such a way that I was taken by a feeling of perplexity as I read the official description for the first time. I was actually astonished. This even made me doubt whether such symptoms could really present themselves in a person who isn't affected by electronic means, such as the V2K. It seemed like some definitions had been extracted from people attacked by psychotronic weapons. In some cases, the symptoms are not only alike; they are exactly the same. I highlighted in bold the most notorious ones in the text below.

We're going to use the ICD-10 Classification of Mental and Behavioral Disorders.

- **Schizophrenia:**

Expression in words that are sometimes incomprehensible. Interruptions and interpolations in the course of thought are frequent and thoughts may appear to be withdrawn by an outside agent.

Ambivalence and disturbance of volition may appear as inertia, negativism and stupor. Catatonia may be present. The beginning of the disease may be acute, with troubling or insidious behavior, strange ideas and conduct.

* Thought echo, thought insertion or withdrawal, **thought broadcasting;**
* **Hallucinatory voices commenting on the "patient's" behavior or discussing among them about the patient or other types of hallucinatory voices coming from some part of the body;**
* Persistent delusions of other types that are culturally inappropriate and completely "impossible", such as political or religious identities or **superhuman powers, as well as communicating with aliens from other planets;**
* Persistent hallucinations of other kinds, when accompanied by "superficial" delusions, persistent overvalued ideas or when they occur every day for weeks, months or years;
* **The most intimate thoughts, feelings and acts are felt as known or shared by others, explanatory delusions, natural or supernatural forces that work to influence the thinking and actions of individuals;**
* Sees oneself as the pivot of everything that happens;
* Perception is extremely disturbed in other ways: overly vivid colors and sounds, and irrelevant aspects of ordinary things may seem more important than they really are.

All of these symptoms are observed over the course of long-term torture. The causes are based on the complete hacking of the mind, mainly the auditory hallucinations. The microwave voice is a factory that creates "schizophrenic freaks". Short interaction of these voices with the target's silent thought is enough for all these symptoms to be detected. Twenty-four-hour surveillance within the person's mind and their habits at home leads to:

- **F20.0 Paranoid schizophrenia:**

More common in many parts of the world, the clinical condition is dominated by relatively stable, often paranoid delusions, generally accompanied by hallucinations, particularly of the auditory variety, and perceptual disturbances. However, the most common symptoms are the ones that left me most astonished, as they are so similar to the effects of psychotronic weapons. In fact, the effects on the brain and mind are identical. Some of them are:

* Delusion of persecution, reference, important ancestry, special mission, bodily changes or jealousy;

* **Hallucinatory voices that threaten the patient or give them orders, or auditory hallucinations without verbal content, such as whistles, buzzing sounds or laughter;**

* Olfactory or gustatory hallucinations, bodily hallucinations (often of a sexual nature), visual hallucinations may rarely occur and aren't predominant;

* **Mood disorders, thought disorder, irritability and sudden anger. Fears and suspicions. Negative symptoms;**

* They may be partial or chronic symptoms. In chronic cases, flowery symptoms persist for years;

* **Delusions of almost any kind, but delusions of control, influence or passivity, or even persecutory delusions of various types, are the most characteristic ones;**

* **Schizophrenic thoughts and peripheral and irrelevant aspects of a total concept that are inhibited in mental activity, usually targeted, are brought to the foreground;**

* It's a form of schizophrenia that seems to come on very suddenly. The average age of onset is late adolescence to early adulthood, usually between the ages of 18 to 30.

* Very dangerous and aggressive impulses may manifest themselves during periods of agitation;

* It's characterized by a lack of activity and response to other people, rigidity of posture **and strange facial expressions (e.g., staring into space and grimacing).**

After a certain period of electronic captivity, if a few hours of their day are recorded on camera, the target will notice that they have characteristics and traits assigned to street freaks. The victim won't be able to recognize their own behavior, since they will be completely different from before the attacks started. As we've seen, the individual now spends 24 hours vocalizing their silent thought with such ferocity to respond to the attackers that they gesticulate, grimace, stare into space, smile randomly, and seem to strike up a conversation with someone or something.

A more restrained person may sporadically scream, in which the projection of the vocalization of thought occurs involuntarily; the more extroverted ones may scream time and again. But, in fact, as their brain is connected or "linked" to the MKTECH system, their thoughts are being sent via Electronic Mind Reading (EMR). They're fighting battles within their own minds against several cowards of this scheme.

* **Persecutory delusion:** belief that someone is chasing and watching you, while planning to do something to harm you. During this phase, the person exhibits behavioral changes, high levels of anxiety and impulses of aggression.

* **Lack of mental fitness**: lack of motivation, apathy, social isolation. The thought fades away and the individual shows complete emotional indifference. Unpredictable behavior, mannerisms are common, haughty postures, grimaces, smug smiles, repetition of phrases and horseplay. Disorganized thoughts and rambling and inconsistent speeches, tendency to remain lonely, loss of volition.

- 20.2 Catatonic schizophrenia:

Constrained attitudes and postures may be maintained for long periods, episodes of violent excitement may be a striking feature of the condition. **It may be combined with a dream-like state with vivid scenic hallucinations**. Perseveration of words or phrases. Inappropriate agitation and posture.

- **20.1 Hebephrenic schizophrenia:**

The disorder is characterized by regressive phenomena such as passivity and personality breakdown. Nonsense and childish mannerisms.

- **F21 Schizotypal disorder:**
    * Eccentric behavior;
    * Anomalies of thinking;
    * Cold or inappropriate affect;
    * Odd behavior;
    * **Tendency to social withdrawal, poor relationship with others;**
    * Strange beliefs or magical thinking;
    * Paranoid or bizarre ideas;
    * Obsessive ruminations;
    * Unusual perceptual experiences or delusions with no external evidence;
    * **Intense illusions, auditory or other hallucinations.**

These are the classic symptoms of schizophrenia. Their great similarity with the effects of this technology — which in some cases are even identical — leads me to profound questions. For example, how can the brain randomly create an eternal conversation with a cadence, fluidity and interaction that is only possible through the relation between two or more humans, including involving different characters, children's voices, adults, robots, etc.?

I propose a challenge to those who suffer from this condition. Enter a shielded room and tell me if you still hear the voices interacting with each other within your mind. See if they talk, have distinct personalities, and if the voices have human characteristics and use words to communicate.

I recently came across a news story that immediately caught my attention. I'd like to share it with the reader after having carefully read the first chapters of this book. Read it and reflect on it. See how thousands of people are victims of experiments with psychotronic weapons without even realizing it. Obviously, there are people with mental illness whose symptoms resemble all the effects we've seen and that have nothing to do with the attacks. However, I feel it's vital to reiterate the need for further studies on people who claim to hear voices in their heads, as such studies are urgent and extremely necessary.

I'm going to highlight the important points that are relevant, fully compatible with the effects of these neural weapons on the human brain and their consequences, as well as the total ignorance about them.

The following article shows that these are the consequences of the involvement of human beings as subjects of illicit experimentation.

The BBC's report on January 29, 2018 writes about Rachel Waddingham who lives with more than five voices in her head that comment on her life on a daily basis. They have different names, personalities and ages. The British woman is perfectly able to distinguish which is which, even by the way they speak.

At just three years old, Blue is the youngest and is very sad, but also very mischievous. Elfie is 12 and is easily offended. Also, since she was 18, **Rachel has listened to three men in her head commenting and criticizing everything she does, as if they were scientists in an experiment watching how she acts.**

"Rachel is stupid", "You're worthless", "Why don't you kill yourself?", and even "You're disgusting, I can't stand you" are some of the things they usually say.

"I remember the first time I heard them. I was in bed and I shivered. I felt I couldn't move and I heard them all saying these horrible things ", she told the BBC.

According to Rachel, the voices also hear each other and speak to one another, some are even afraid of others. **"It's like having a network of people in your head."**

Doctor Angela Woods from the University of Durham, United Kingdom, conducts the project "Hearing the voice", one of the most complex studies on the experience of hearing voices. She says that most people experience some auditory illusion at some point in life, especially during sleep or, for example, when you think someone said your name. It's estimated that 2% of the population hear voices that "live" in their heads. **"These voices are as real as the things we experience in the world"**, Woods explains.

The report also talks to other patients with the same "illness". One of them, for example, wrote a book for children thanks to the "collaboration" of one of their voices.

Patients say that voices are often critical, but there are others that are friendly and even give advice.

Different characters who live in the same mind can also speak different languages. They're usually beyond the control of the patients, although some of them say they manage to control the voices in a certain way.

For some, these voices are in no way different from those we hear from people in the real world. **For others, it's like constantly listening (in the background) to a conversation at the next table in a restaurant.**

Rachel says that sometimes she notices the presence of the voices, even though they say nothing. It'd be almost like a sensory experience, in addition to the voice.

She was diagnosed with a combination of schizophrenia and bipolar disorder. She got into a routine of medications for years, with the use of antipsychotics and frequent visits to the hospital.

For the doctor, this phenomenon is linked to the trauma experienced in childhood.

The explanation really convinced the patient who states: "I believe I silenced the abuse and reduced it, and the voices are almost like metaphors or windows for that. But, is that the reason I hear voices, because of the trauma? I don't know. Maybe it's because I genetically have the ability to hear voices".

Retrieved from
http://www.bbc.com/portuguese/geral-42827481

※ ☼

Given what we know — based only on the first chapters of the book — what conclusions can you draw from this report? If you are still in doubt, we shall continue talking about it. In the next chapters, we're going to go even deeper into a whole new world in search of total mind control. This will change the way we see reality. And as for the news, to get to the root of the problem is simple: instead of treating the symptoms, we must treat the disease called MKTECH and its system divisions specialized in hacking the human brain.

CHAPTER 2.4

EMRA – ELECTRONIC MIND READING (AUDITORY)

"I was listening to a song at full volume using a headset, loud enough to damage my hearing, and even so I continued to hear the voices that seemed to come from within the song, singing the lyrics and replicating part of what I thought silently! Something out of this world!"

— **Anonymous Targeted Individual realizing that the voices inside their mind weren't related to a disorder.**

As indicated in the chart above, we're gradually advancing and diving deeper and deeper into the bowels of this technology. We now begin to correlate its functions and modules to cognitive processes. The brain has some very specific and quite independent cortical mechanisms. Each MKTECH module or subsystem is responsible for interacting with one or more of these mechanisms. Such mechanisms are similar to computer systems as we know them today, subdivided into interconnected modules that interact with each other and create a single absolute integrated system.

So, in order to explain what V2K (Intracranial Voice) is and how it works, we had to take a quick look at how the human being captures the sound, the complex processes involved in sensory interpretation and how this capture affects us psychologically and physically, along with the great ability to change our behavior. Now, we're going to know a new powerful remote invasion system, a subsystem within the mind control technology. This module is called EMRa – Electronic Mind Reading (auditory), a system capable of "listening" to everything the target is hearing. Everything that is captured by the ear and processed before being emitted to the auditory cortex coming from the afferent pathways (hearing) is amplified and sent to the MKTECH computers, becoming a kind of tap, however, executed completely remotely and using only the target's auditory circuits as equipment. This is another module of this surreal electromagnetic "non-lethal" weapon, capable of serious violations against humanity.

Yes, reader, it's possible to listen to everything that the person connected to the MKTECH technology hears — wherever they are — without the need to install on site any type of equipment commonly used for this purpose. From the simplest ones, such as cell phone microphones, recorders, wires and parabolic microphones that capture sounds at a certain distance, to the most complex, which emit laser beams that register the vibration of the window they're aiming at, making the sound that reverberates in the environment be heard, among other electronic spy equipment. None of this is necessary, just a series of electromagnetic signals that act directly on the target's auditory system, and that can be executed and sent hundreds or thousands of miles away from the mapped brain, given the natural characteristics of electromagnetic waves which have an edge over any other equipment listed above. Consequently, this system can be performed in total silence [22]. The target will never know that the sounds picked up by their ear and interpreted by the cortex responsible for hearing are amplified and diverted to the invading system.

It's worth remembering that MKTECH spends 24 hours a day stealing the target's vocalized and visual thoughts, as we've seen in the previous chapters. Now, another very serious parameter that is

---

[22] - In the initial connection of the MKTECH system to the mind of the person who will become a target, however, a characteristic ringing in the ear is periodically felt, and this buzzing becomes more and more acute until it disappears.

going to be covered in this chapter in detail is also captured by this technology: all ambient audio that will be processed by the victim's auditory cortex. That is, technology operators have unrestricted access to their conversations in any place and circumstance of their everyday life and, mainly, their private routine.

In order to have a notion of the amount of information that can be captured and extracted from the target's auditory circuits in a short period and during their daily activities, we're going to make a new contemplative pause. It's going to be different from the initial chapter, in which we closed our eyes and focused on a mental image of a house to access visual thoughts.

Let's stop for a minute now. Close your eyes to focus your attention on a cortical area of primary reception. Focus only on what you're listening to. Pay attention to the amount of sound stimuli around you, to the quality and intensity of each sound wave reaching your ears, natural sounds from the environment around you, from the TV, noise from cars on the street, the wind in the distance, footsteps from people in the apartment above, voices of passers-by on the sidewalk, breathing sounds, trucks, motorbikes, the creak of a chair or a cell phone ringing. Ready?

The cacophony and external excitement that generate the *soup* of noises and sounds that you're hearing at the moment are mechanical sound waves, traveling through the air and being picked up by your auditory system. Now think: even counterintuitively, that one step before being interpreted, of having a cerebral representation of the sensory event in your mind, all these different sounds and noises are amplified, diverted and acquired by the computers of the MKTECH systems using the Electronic Mind Reading (auditory). Look at the convenience, ease and efficiency of this system to capture data without using any other device foreign to the environment.

In addition to completely violating the privacy of the Targeted Individual, people who are directly interacting with the victim are also involved and the privacy of third parties are violated as well. The severity of this invasion gradually emerges with each interaction with other people. These day-to-day interactions generate content which can be of any nature — from a simple conversation about current events, intimate conversations with affective relationships or friends that may recall childhood issues, conversations of a family or school nature to subjects that have a meaningful context only to the people involved, but that can directly affect everyone. They can access memories full of positive emotions, evoke memories modulated by painful feelings, or it may be a subject that will be remembered and forgotten right there.

A casual encounter can be captured and evaluated by operators to then be used in attacks via Synthetic Electronic Telepathy (SYNTELE), as also happens with subjects in situations considered ordinary and unimportant by "normal" people who don't have their brains kidnapped by MKTECH. However, the situation worsens when the conversation is sensitive, whether of economic or commercial relevance, potential intellectual property, confidential matter that exposes secret or creative products to be stolen and patented. In other words, any theme or idea that may be of value is absorbed, captured and stored by MKTECH.

So, we have two distinct and associated purposes in the use of this tool. The first is of a commercial intent, of theft of ideas and products. The second has a highly nefarious and disastrous purpose: to solely collect information about the target's intimacy — their social circle — to be used as ammunition in subsequent attacks. Personal information of close relatives captured in conversations may be used to attack, threaten and create an atmosphere of superiority, as it morally reduces the target and causes hesitation and doubts about the success of any reactive action against the tortures. By spreading extremely personal information about the target's family, a unique ability to control the entire environment is demonstrated. They then start to verify the interaction between the target and the world around them, mainly in the social aspect. People who know nothing about MKTECH are involuntarily involved in this obscure world.

There is a program in the telephony and telematics area that acts in a similar way to EMRa. This program is called Guardian and is used by intelligence and security authorities. It automatically taps every phone that makes contact with the main phone object of the investigation. Regarding EMRa, what is being "bugged" is the target's nervous system. These waves that amplify and interact with the brain cause physical and psychological damage in mammals over time, even if they aren't exposed to radiation for a long time.

If the operator eventually overhears a conversation that is relevant to their objectives, the interlocutor of that conversation interacting with the target will have their brain silently connected to the scheme due to the EMR. A parallel recording process will be initiated and will store and select the relevance of the subject without human interference. Subsequently, a computational algorithm will divide the subjects based on parameters pre-configured by operators, until it succeeds in extracting some subject that can be used against the target, or that has commercial relevance.

The purpose of SYNTELE torture is to maintain the constant stress caused by a cognitive violation, as it makes "noise in the auditory cortex" of the target which, in itself, is completely maddening. However, the content of this noise is what really initiates the process of physiological alteration. Have you ever heard the phrase: "Words have power"? This statement is not for nothing. And this repertoire of words that have to do with the target's reality must constantly be renewed in order to affect the individual in an overwhelming way at all times. After all, some of them lose efficiency as the target recovers from these terrifying attacks.

Listening to personal conversations is a great strategy to raise deep questions and reflections on subjects related to the victims. It triggers ongoing fears of seeing their loved ones suffer physical violence, and it reveals emotional vulnerabilities and a range of feelings used to renew the arsenal of attacks to be carried out in the future. Therefore, this information is of great importance. The operators carrying out the attacks speak openly to the target, follow their routine, easily invade their privacy, know all their secrets and listen to family and intimate conversations. Consequently, the target is aware that they're listening to such interactions. The situation, as a whole, causes an anticipation that something unpleasant may happen with someone they know, revealing more circumstantial details that serve as ammunition for the observers. There is also the discomfort of being aware that absolutely all aspects related to social interactions are being observed, raising questions about how the tortures will use certain conversations for their own ends.

Over time, the victim increases the escape response in certain social interactions of everyday life. They deliberately isolate themselves so as not to create a chain of future attacks against the people who interact with them, which makes social relations more difficult. This obviously depends a lot on the psychological profile and on how all this cognitive violation negatively affects the target. Keep in mind that those who interact with the victim will act normally, just like any person would do. They won't restrain themselves or moderate their words as they're unaware of the invasive presence of the operators who are "installed" as parasites in the auditory pathways.

These day-to-day conversations and interactions also serve a very peculiar, surprising and gloomy end: providing the operators' voice database with voices familiar to the victim. The ambient sound of the places where the victim usually goes for a walk is also recorded, as well as interactions and conversations. But for what purpose do operators record this? For immersion in the dream soap opera and storytelling used to contextualize a certain situation within the target's dreams, as they will react spontaneously when hearing familiar voices during their sleep and in the course of the attacks via D2K (Synthetic Electronic Dream or Dream to Skull). In the next chapter, you're going to be introduced to the most surreal weapon in the MKTECH system.

Other aspects, that are studied and collected, start from the following premise: almost all human behaviors are the result of coexisting with others. Social interactions and social behavior result from the environment that shapes us. From there, we see the magnitude of this MKTECH module working together with others.

Operators can monitor the target's reaction to external sensory stimuli wherever they are, detecting the slightest physiological change, and how they interact in certain conversations with their friends and acquaintances. In addition to listening to the subject in discussion, the EMRvi – Electronic Mind Reading (vocalized/images) remains constantly connected in parallel with the EMRa – Electronic Mind Reading (auditory). That is, they monitor all daily conversations, checking how the target reacts to every situation in the most varied interactions and in the topics discussed with the most diverse people. The verbally expressed responses and the reaction to each interaction are joined together with the silent thoughts that naturally arise in the target's mind in response to the main stimuli resulting from such interactions. In this way, the theme that evokes memories of any kind — which makes the target uncomfortable, worried, or causes any emotions — changes with each reaction to the issues raised, generating restlessness, happiness, calm or sadness. It helps in obtaining a lot of personal information and a more refined emotional and psychological picture of the target.

They can now use the intimacy of family members and their exchanges with siblings and parents, uncles and cousins or even close friends to capture possible memories and unfavorable everyday situations, family problems from the past, or any relevant information that is absorbed in direct conversations or in thoughts of the target's family members on a given matter. The victim may even be surprised with some early news of important events relevant to the family before being officially informed by the person in charge who was preparing to announce the fact — perhaps with a small formality or during a casual message conversation ("Hi, we need to talk"). The topics can be varied, such as breakups, something related to unemployment, illnesses, death, changes, and so on, and they can be negative or positive, since all close people's thoughts are easily captured and monitored.

To illustrate my point, look at this situation: the target's girlfriend indicates she is going to end the relationship or is interested in someone else. This information is detected by the operators and sent to the target's mind, which anticipates the events. This fact is then transformed into emotion-laden content and is used as ammunition. It can even make the targeted individual commit crimes of passion, or even worse crimes. It's not enough to just make noise in the auditory cortex; they use deep psychological factors inherent to every human being to perpetuate the torture and theft of information. In addition to this advanced information gathering, there are also degrees of psychophysiological electronic warfare tactics used in the targets that we're going to see soon.

Along with listening to what the target is hearing at the moment, all other cortical areas are being monitored. That way, operators can capture reactions to these conversations and day-to-day interactions in several ways as they listen to their internal thinking, visual data, physiological reactions, etc. Therefore, they form a complete picture of reaction and interaction with each word spoken in conversations. This interaction results in a mass of data that would be impossible to use efficiently without intelligent and dynamic filtering. Listening to all the target's conversations and sounds and their subvocalized thoughts requires advanced algorithms to filter what doesn't matter and what does.

A huge amount of cognitive data is produced each day during mass capture. To handle this information in different formats, they use the concept of NEURO BIG DATA [23] and AI (Artificial Intelligence). The volume is filtered by an autonomous program using the machine learning concept — neural networks —, which can be pre-configured by the operators with some parameters indicating to the algorithms some basic premises to, from there, start the data filtering, in order to evaluate the relevance of the subject with the manual selections. For example, the algorithm can be trained to compare EMRa data in real time. It searches for certain keywords, a place with a "party", a name, dates, or actions such as "sold" or "died". So, as soon as these words are captured, the algorithm points out that a specific word has been detected, and that part of the conversation is now placed in a location where there are conversations considered relevant. At the end of the chapter, more details about the screening system are going to be provided. The book discusses in more detail the algorithms used and the computational systems of this technology in chapter 11 of volume 2.

There are also some auxiliary algorithms which estimate the distance at which each sound is captured by the target. They indicate to operators how many inches, feet or miles the source emitted a certain sound that was picked up by the ears at a given moment. The greater the intensity and purity of the sound, the closer the object

---

[23] - "Big Data" is a term widely used to name very large or complex data sets, which traditional data processing applications cannot yet handle. This term often refers to the use of predictive analytics and some other advanced methods to extract value from data, and rarely to a given size of the data set. Greater accuracy in the data can lead to more confident decision-making. "NEURO BIG DATA" has the same functionality as big data, but only manipulates data processed directly by the brain.

that produced it is. This is yet another example of the amount of information that ends up with the operators in real time. These data have the total control of all events that interact with the victim and the dynamics of the environment that surrounds them.

But how does it happen?

In the early days of research on neural weapons — periods that we're going to experience in an immersive and shocking way in the chapters ahead — the people involved had an epiphany during some experiments and ended up using an elegant solution to resolve this standoff. Where does "raw" data go before it is processed by specialized cortexes? Where is this electrical signal that carries information "channeled"?

The vast majority of scientists who research the brain usually immerse themselves in the data and try to trace paths through the neural circuits to detect which information captured stimulates a certain area in the cortex. The scientists who developed this range of psychotronic weapons took a different path, which in fact proved to be a very ingenious technique. After all, it's not necessary to trace the ways in which thoughts are sculpted by measuring the complex behavior of billions of neurons. Just "amplify" the signal and capture it already electrically prepared to be interpreted by the responsible cortex. Then, concentrate efforts on this common bridge through a central channel responsible for the delivery and where the information converges and is concentrated. This place is called THALAMUS.

2.4.1 - Thalamus

The thalamus — with the hypothalamus — forms the diencephalon, which are located in the center between the cerebral hemispheres and the brain stem. The hypothalamus contains centers for many functions, such as thermoregulation and water balance, and it also sets the feeling of running or fighting and visceral reflexes. However, our focus is the thalamus, better known as the **gate to the cortex**, where eyes, ears and skin establish synapses of retransmission before reaching the Central Nervous System. So, the thalamus is of great importance, as it processes almost all sensory information that reaches the cerebral cortex, except for smell.

The thalamus and afferents to the entire Central Nervous System can act as a powerful *pacemaker*. Under certain circumstances, they can generate very rhythmic action potentials. The thalamic cells present a set of voltage-dependent ion channels. They synchronize the rhythm of the group of afferent neurons, functioning as an oscillator whose main function is to serve as a reorganization and integrating station for stimuli coming from the periphery, as it uses the Relay Core [24] to integrate converging information. These nuclei are especially prominent in the thalamus. In the sensory receptors, such as sight and hearing, the pathways that take different types of information to the central nervous system begin, each one of them carrying a specific type of message. In the case of the auditory system, the external sound information is converted into electrical impulses after being processed by specialized cells until it reaches the auditory cortex and is interpreted. The concept also applies to sight.

Almost all of the main pathways that go to the cerebral cortex have a synaptic station in the thalamus that is predominantly made of subcortical gray substance divided into several nuclei, each of which receives a type of afference and projects it to a specific region of the cerebral isocortex. Visceral functions, sensory stimuli, emotional behavior, touch, somatic sensitivity of part of the head, somatic motor skills, sight, sound, even language has a direct connection with the thalamus.

The thalamus also serves as an oscillator that helps to control the state of sleep and wakefulness. In relation to hearing, the Medial Geniculate Nucleus (MGN), located in the thalamus, receives information from the cochleas — as we've seen in the chapter on V2K. At this point, the sound has already been transformed from a mechanical wave into electrical stimuli that will be "delivered" already processed to the Medial Geniculate Nucleus by the arrangement called acoustic radiation to later be distributed among the various cortical

---

[24] - A relay is an electromechanical device with numerous possible applications in switching electrical contacts, as it serves to switch devices on or off. It's normal for the relay to be connected to two electrical circuits.

areas specialized in the interpretation of sound and its particularities, such as the auditory cortex. There is ample feedback in auditory pathways, specifically between the auditory cortex and the MGN.

It was then that the scientists working on the MK-ULTRA program (Volume 2, chapter 4), with the task of creating a method of capturing everything the target listened to, developed a genuine Neural Sniffer [25]. As soon as the information converges through the nerve pathways until it reaches the Medial Geniculate Nucleus (MGN), the signal is amplified, captured, modified and loaded into another carrier wave that is sent to adjacent antennas and retransmitted to the remote base where operators (thieves, leeches) are located. There, a complex program, a Brain-Computer Interface (BCI), decodes this signal in a similar way as the auditory cortex would do in the brain, interpreting this information and acquiring the ability to hear everything the target's ears capture, that is, all sound waves already converted into impulses, including vocalized thoughts that are being processed.

Thus, there is the EMRv - Electronic Mind Reading (vocalized), which converts the captured signals into transcribed texts based on the capture of audio from the vocal cords before they become the speech motor processes. The EMRa – Electronic Mind Reading (auditory) turns vocalized thought into sound, and becomes a redundant, practically infallible system. This is one of the reasons why targets never stop having their thoughts stolen and why V2K torture never stops, wherever it goes. This is a vital module of MKTECH, unlikely to fail.

This technique is the "simplest" way to amplify the signal and decode it later, since the MGN is a centralized channel that transmits the electrical impulses already "chewed", ready to be interpreted and refined, to the auditory cortex. It becomes much more difficult to have control of the information and to amplify signals that are already distributed throughout the specialized cortex — and increasingly centralized neurons — in places where functions are very scattered, with each neural interaction. For this reason, the thalamus is used as a data capture center — data diversion — or neural hacking.

---

[25] - "Sniffer" is a computer program that captures all data packets that travel over a given network, including passwords and unencrypted usernames. It works in a similar way to the MKTECH EMRs.

The era of wiretapping by using electronic equipment with hidden microphones, or directional microphones that record from a distance, is a museum relic, even though it came well after MKTECH.

CHAPTER 2.4 – EMRA – ELECTRONIC MIND READING (AUDITORY)

Process of capturing audio directly from the victim's brain that is connected to the system:

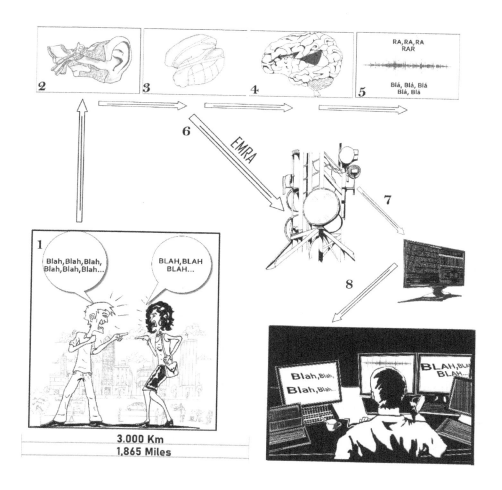

Figure 2.12 *Steps 1 to 5 are the natural sound transduction processes. Steps 6 to 8 are the processes of capturing the sound content directly from the target's mind.*

1) The target argues with someone in any hypothetical everyday situation. The vocal cords vibrate generating audible variations in air pressure, the sound.

2) The sound (mechanical wave) is picked up by the ear. Processing of mechanical waves into electrical signals sent to the thalamus.

3) Thalamus, more specifically the Medial Geniculate Nucleus (MGN), which is responsible for transferring the signal to the auditory cortex where the content of previously spoken words travels centrally.

4) The auditory cortex makes sense of the words captured by processing and interpreting electrical signals. Brain activity area highlighted in black.

5) Final meaning of the sound picked up by the ear; how the brain understands meaning when uttered in words.

6) The moment the information travels through the MGN (step 3), an electromagnetic signal sent by antennas or satellites "amplifies" that information, which is picked up by other antennas. The signal is "diverted" to the system.

7) The signal is then routed to BCI algorithms where EMRa subsystems translate this signal analogously to the auditory cortex.

8) The translated signal is converted into a comprehensive language for human beings (sounds and automatic transcriptions) and transferred to the operators' computers — the thieves, hackers of the mind.

Vocalized thoughts

As previously mentioned, EMRa also serves as a redundant system for capturing the vocalization of thoughts. Remember: for the person to generate vocalized thoughts, they have to listen to this vocalization of thought, in an integrated system. Thus, the target's auditory system needs to be intact and in perfect working order.

The vocalized thought travels through the same auditory pathways. It goes through the MGN, in the thalamus, like any sound captured externally, despite the vocalized thought being generated by internal processes, since there are cells that respond to very complex sounds, such as the vocalization of thought. It becomes a second capture route used by operators and, in certain situations, it can become the main route of capturing vocalized thoughts. As soon as the vocalized thought is being automatically generated, it will go through the auditory processing in a constant and involuntary cycle. If it didn't, you wouldn't be able to hear your vocalized thoughts, nor the content of the book you're currently reading!

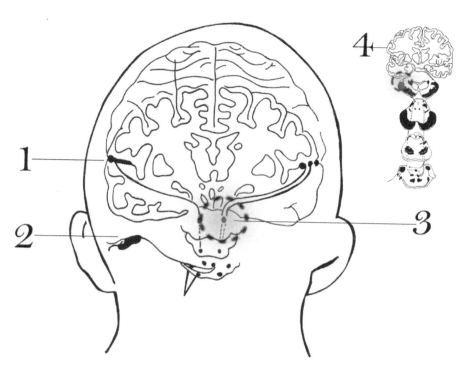

Figure 2.13 *Brain image highlighting the exact point of the MGN or Medial Geniculate Nucleus.*

1) The human auditory cortex is a defined region that constitutes 8% of the total surface of the cerebral cortex.

2) Cochlea — the cochlea, or snail, represents the auditory part of the inner ear located in the temporal bone. The cochlea interacts with the middle ear via two holes that are closed by membranes.

3) Medial Geniculate Nucleus is a structure composed of several nuclei with a very complex and important functional role. It's believed to be the exact point at which the EMRa re-radiation "captures" the auditory data and modulates its return wave, including the target's vocalized thoughts and cognitive privacy. Your neurons continue — in the auditory pathway — to send their upward projections to the primary auditory cortex. When we think in words, we hear our own silent thoughts. These thoughts are processed by the auditory cortex and travel through the MGN. Thus, they can be captured via electromagnetic re-radiation.

4) Layout of the cerebral cortex highlighting the midbrain, the Medial Geniculate Nucleus and other structures.

V2K microwave voice capture

The EMRa also has an important characteristic for MKTECH technology operators: it verifies if the microwave voices are hitting the target, if they're being interpreted correctly, their degree of intelligibility and if they're being picked up by the target without any loss. This confirmation takes place in two ways: by the interpretation of the V2K content itself, in which the operator is able to capture the audio that is being sent and thus confirm that it was correctly interpreted as a sound that came through mechanical waves; or by the confirmation of the activation of the target's own vocalized thought, which is involuntarily triggered in response to the stimulus in the content interpreted by the intracranial voice.

Every MKTECH module shocks us more and more with each chapter read. Nonetheless, don't be surprised! It'll get better — or worse, depending on the point of view. The thalamus hacking created a secure channel for practically all the important data to be captured via EMRs, including the sight, which is going to be explored later on in the book.

2.4.2 - How the algorithm that filters conversations works

As we've seen, depending on the group behind the MKTECH scheme, artificial intelligence works in the valuation decisions expressed in the messages.

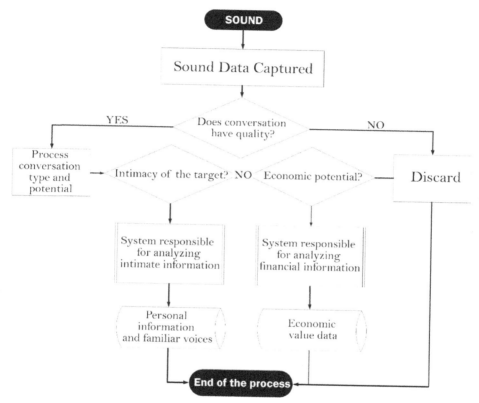

Figure 2.14 *Algorithm responsible for filtering the content of conversations.*

CHAPTER 2.5

D2K – SYNTHETIC ELECTRONIC DREAM

"You running and you running... away. But you can't run away from yourself."

— *"Running Away"*, Bob Marley.

The next system to be discussed seems to have come out of science fiction movies — just like the previous modules. This is yet another non-invasive form of torture, theft of information and total violation of the privacy of others in the sense of physical invasion of the organism. The Synthetic Electronic Dream, also known as Dream to Skull or D2K, is a weapon with the very specific purpose of completely mastering cognitive processes while the target is unconscious (sleeping). This weapon was classified as Top Secret (the highest classification level) when it was developed in a MK-ULTRA subproject codenamed **"SLEEPING BEAUTY"**. Its purpose was to modify memories and alter the content of dreams using radio frequency (chapter 4 "MK-ULTRA", volume 2). It was never heard of again, until it began to be used for torture and cognitive experiments with MKULTRA 2.0 (volume 2, chapter 5) in modern times. Within the set of systems, and in the sense of violation and aggressiveness, this is undoubtedly the most dangerous, invasive, destructive and powerful weapon of the MKTECH system. It opens up unprecedented possibilities of total control over the waking human mind, as they do when the target is asleep, unconscious. They practically have complete control over the graphic content that will be displayed within the individual's mind.

Welcome to the darkest module in all Mind Control Technology system!

This subsystem can be summarized as a powerful electromagnetic weapon capable of completely replacing a person's natural dream with electronic dreams artificially generated by technology operators. They're transmitted in a similar way to an analog TV, with images and sounds in different frequencies. However, they're demodulated and interpreted by the brain, not the TV, causing vivid and extremely real dreams every night that can only be compared to the worst nightmares most human beings experience sporadically in their lives.

The natural dream, which would normally be created by the individual's mind, is replaced by these prefabricated transmissions. This technology is able to emulate the frequency in which the dream is demodulated and interpreted by the areas responsible for presenting the final result of the dream. It thus replaces the natural signal — that would normally be created by the content of the brain's electrical storm — by an artificial transmission. This leads the visual cortex to interpret these invasive images, triggering the whole process of a normal dream, however, based on artificial content, which reflects directly on the target's entire nervous system and forces them to dream the content that the operators are transmitting at the time.

That's right! It's possible to completely replace a human being's dream with any content that operators want. This was inconceivable in the past — perhaps it existed only in movies. Now it's a reality that comes to reinforce the fragility of the mind and its processes that are easily hacked! Consequently, dreams of any kind can be generated with any content.

There are several techniques that can be employed to generate the content of the Synthetic Electronic Dream. This also depends on the operators' goal at the moment, since the dream has a feature known by few — and because it's only possible to effectively verify this with this technology. For those who don't know, **R.E.M** stands for "Rapid Eye Movement", the phase of rapid eye movements in which vivid dreams occur. The way a person — who is sleeping soundly and falling into REM sleep stage — interacts, the responses given unconsciously within certain situations imposed on the dream provide a lot of relevant information about that person. The individual cannot hide anything related to their personality within the dream. They cannot pretend or lie, because, in general, the intellect only partially falls asleep. Most of it remains active during the dream.

The person in the dream is actually the projection of himself or herself within another context of reality — a reality that only exists inside their own mind. This content can be remotely manipulated during the occurrence of the dream. However, the brain interprets it as the reality that we live in a waking state (awake) with some mitigation and suppression, such as the sensation of pain that doesn't pass from the reality of the dream to the physical body, and the reduced sharpness of the interpreted images. That is why this is the perfect tool for testing situations, "creating probabilistic models" with the targets to see their likely reaction outside the dream — in concrete reality — if it were to happen. The target within the dream only reacts to the stimulation of the environment as they would in similar scenarios when awake. The results will then be pointed out as a likely reaction to the same situation if it happened in the common reality. Therefore, D2K is extremely useful for simulations with the Targeted Individual and also for discovering their most intimate secrets!

Generally, Mind Control Technology operators have a soundproof recording studio furnished with professional equipment that includes cameras, chroma-key, microphones, computers, costumes of animals or of characters from fables, toys, miniatures, modeling clay, 3d and virtual glasses, among others. A real scenic area, a place with equipment used in theater, such as lights, a sound table, big screens to see what is being filmed (what is being transmitted as a result of a montage), and a screen with the target's return data, that is, their reactions to the scenes inserted in the dreams. In addition to all the apparatus, the performances are carried out every night based on a different script, a new dramaturgy, with fictional characters and human actors, in a *soap opera* similar to those from the 90s and early 2000s. The characters of this dreamlike drama will accompany the target for years to come.

The operators compile the result of the daily filming done using a regular video recording camera, edit the film with video editing programs and send these sequences of images by means of electromagnetic waves to the brain of the mapped individual, which makes it possible to interact with the victim in REM sleep in real time. That's right! It's possible to receive the broadcast while sleeping. Operators play with the targets' brains, as they completely modify the natural content of the dream. Thus, both pre-recorded and live content can be transmitted, which makes D2K a vital complement to the daytime torture undertaken by SYNTELE. However, something more obscure than simply unconnected changes in the content of dreams hovers in the air. In fact, this is the moment when most of the experiments conducted by MKULTRA 2.0 (chapter 5) are tested, as we're going to see during the course of this chapter. The same techniques used in films are used to create more sophisticated and realistic remote dreams.

At the end of dawn and early in the morning is when re-entering the REM stage becomes faster and the dream has an extended duration. MKTECH is capable of conducting a complex narrative similar to movies and soap operas, as it takes advantage of this natural brain event of vivid and extended dreams, in which the protagonist is the dreamer who navigates through various scenes and faces different characters. These characters build interactions and deep and complex debates involving the target's life. They create unprecedented situations, recreate past events, redo the scene and question every decision the unconscious "self" made in reaction to imposed scenes. It even creates future plausible projections about the reality that surrounds the victim.

In addition to creating a complete psychological picture of the person, practically all secrets that orbit a given context of an event are revealed in small details. This, in turn, provides an enormous amount of data and at the same time alters the perception of reality, as it creates lasting long-term memories, mental confusion and the infamous "dissonant mind" or "dislocated thought", which is going to be discussed more deeply later on.

The D2K is transmitted to the victim's unique frequency, otherwise, everyone around them would have the same dream, which would be extremely weird and would generate distrust among people living in the same area. The operators would indeed compromise the entire scheme. This intrusive film is then assimilated by the brain as if it were a natural dream, and the responses to the stimuli of the dream content are captured via EMR (Electronic Mind Reading) and remote EEG. They use the Intracranial Voice (V2K) for auditory communication. Operators interact with the victim in real time in a lifecycle that resembles that of computer games. It causes the individual to unconsciously interact with the false dream, however, this interaction generates a series of responses that are captured by computers that, in turn, resend a new situation in the dream so that the victim remains immersed in that dream-game (film or dream-film) reality presented by the operators. This creates more spontaneous responses and allows them to go through stressful situations, since we have no real experiences in the world, but sensory experiences resulting from stimulation in the cortex both awake and asleep.

Some of these experiences are so striking that they have the capacity to carry negative feelings to the waking state (awake), causing mental confusion in the target for a long time. After all, we all know how a good night's sleep and positive dreams can help us have a more energetic and mentally prepared day. The opposite, on the other hand, makes us start the day feeling sick, with little receptivity to events, low attention and excessive focus on the memories of the bad dream from the night before.

Fully manipulated dreams produce a profound effect, added to the emotional impact associated with the unthinkable — of experiencing something that was thought impossible. In order to understand how huge this impact is on the organism, we first have to consider how the person who has their dreams remotely manipulated feels within the virtual world created by strangers. Generally, all negative emotions are experienced. The films are deliberately created with the intention of harming, modifying and destroying — as much as possible — the cognitive processes of the target, both awake and asleep. The feeling of having your privacy violated in an even more painful way than SYNTELE does is included in this destructive package.

So, in order to achieve this effect, it's only necessary to make the brain interpret the images during the dream. The rest, emotions and memories, is accessed automatically, as the mechanics of the dream remain the same. Only images and sounds are created electronically, and they superimpose the images that would be generated by the brain. Operators can also resend signals with negative primitive and emotional EEG settings so that the dream becomes more realistic or striking. This causes harmful physiological responses in the human body at the exact moment this attempt to reconfigure the electrical patterns of certain areas of the brain directly involved in dreams occurs. It's highly disturbing to have your dreams remotely manipulated by third parties at night, not only for the act itself, but also for the immersive and interactive capacity of this technology, which can enhance the insertion of false memories. Dreams generated in studio, and with computers, are always extremely vivid and end up confusing the victim about the origin of the memories — if they were acquired in concrete reality when they were awake or if those memories are the result of nightmares or manipulated dreams.

Electronic dreams always have a level of sharpness and vividness far superior to everyday dreams. They're compared to the worst nightmares that a normal person has ever had during their life with the difference that those manipulated via D2K are able to directly affect long-term memory in a completely unprecedented way for humans. This is because operators know exactly the best time to wake up the target from these vibrant, powerful and totally strange dreams. So, the transition to the waking state completely alters the target's life, since dreams are now as or more intense than reality!

In this context, dreams have no philosophical or prescient meaning, connection with other planes or anything in that playful and romantic sense. They serve only as an instrument of torture, brainwashing, negativity, mental persuasion, to discover or confirm details about the victim, and to simulate scenarios. They're used to support the confirmation process of the remote polygraph that we're going to see in chapter 3. They serve as a basis for validating details that occurred in a given moment in the past, to discover the victim's weaknesses and to explore them later, triggering reactions that cause harmful thoughts when the subject is delivered via V2K. They serve as a scientific experiment to test cognitive tools, neural responses, sensations and emotions recruited from each interaction with the victim. They're extremely valuable as a test for improving weapon methodologies and for commercial purposes as well. Any individual affected by this technology is a guinea pig, a victim of experiments that are prohibited by law to be carried out on humans. These experiments are subject to prosecution and restitution due to the hours in which the targets were forced to participate and without pay.

The total manipulation of dreams and the distortion of sleep architecture cause irreparable damage to the target being attacked and studied. Their entire creative process and their freedom to dream naturally are restricted electronically. They can no longer count on sleep and dreams as a form of "renewal". The brain no longer reinforces the relevant facts of the day, as these are completely suppressed and replaced by false images. Associated with this, operators interfere in the brain waves responsible for the person's emotional state, in addition to the natural reaction of the hypothalamus to the images and sounds presented in the electronic dream. This causes automatic reactions of primary emotions, inserting certain configurations related to negativity — anger, fear, discomfort —, so that the electronic dream or nightmare is the worst possible experience for the individual. This causes reverse effects, unlike those from natural dreams (primary function of solidifying memories and cleaning the neuronal communication system). Now, memories of the dream only occupy time and space in the biological hard drive. The target will have to live with these prominent artificial memories for the rest of their life, because both long-term and short-term memories are affected.

This electronic dream replacement tool is extremely powerful. If operators want, they can create intense suffering. However, the unknown side of the system is noticed on rare occasions when the target is subjected to torture and experiments — if it were used for useful and peaceful purposes, of course. There is an immense potential to add knowledge to an individual's brain, transmitting formulas, study techniques or any other content that helps the target in their daily lives. Weeks before a test, for example, it'd be possible to insert content related to the subject, which would assist the learning process since it is possible to insert any type of memory during REM sleep. They can also keep the target in deep sleep as much as possible, making them dream about good things until they wake up naturally for other reasons, such as biological needs. This technology can make you embark on a highly realistic and immersive, non-destructive dream that produces positive feelings and emotions lasting for hours. The brain takes advantage of these rare moments of peace in the target's life and automatically remains in a state of deep sleep for as long as possible. In other words, D2K can be used to make the brain embark on a lasting and invigorating sleep. It can also be used as a scientific research tool on the psychophysiological nature of the sleep state, capable of offering a powerful approach on the perception that surrounds the dream and its mysteries. Unfortunately, the system is a weapon designed to destroy, kill or drive someone crazy.

Another way to use the technology — in a way that it doesn't harm the individual on purpose — is to create an interference in the natural dream, resulting in a null or neutral dream, with no content. It serves to prevent the transmission of the normal dream from being interpreted, not allowing the brain to create and naturally process the dream. It's like a jammer that interferes with the transmission. In this case, only noise and diffuse images or total silence and completely dark images are transmitted.

Imagine that your brain edits a film whose content comes from various different areas. The visual part comes from the visual cortex and the visual memory passing through the Lateral Geniculate Nucleus (LGN) located in the thalamus. The audio would come from the auditory memory and the language of the parts responsible for this system that are spread over several different regions in the brain. Thus, the intruder film arrives inserted in an electromagnetic signal identical to the natural one, in which it is transmitted to the regions of the cortex responsible for the interpretation of this signal. It simulates the information that would come from the regions responsible for sound, image and language through the D2K and V2K devices. Dreams are partly influenced by the auditory cortex, which is one of the active external stimuli. Operators transmit waves that resemble TV waves (AM - VBS) from an external device. The dream that is "edited" by the brain and artificially created by technology operators is vivid, immersive and impressive. It's only necessary to replace the visual part of the dream to generate the effect on the mind. It acts directly on the visual cortex via the thalamus and the specialized nuclei by visual perception. The rest is triggered naturally, just as the emotions that are activated by the situations imposed on the movie scene (dream).

The tremendous power of psychotronic weapons that interfere with electrochemical properties of the brain is verified in the D2K system. When the target is unconscious from sleep, they have all cognitive functions hacked by MKTECH operators — all except sympathetic and parasympathetic autonomic functions. The operators simply turn the brain into a biological data interpreter with devastating consequences for the person whose natural cognitive functions are completely altered and violated. Luckily for mankind, this technology still doesn't have full access to the brain of people who are awake in the same way that it works when they're sleeping. This, however, may soon change with the constant improvement of the technology.

D2K deeply works with the human psyche thanks to very specialized Artificial Intelligences (AIs) that are part of this module. They have precise sensors capable of indicating which stage of consciousness the target is in, pointing out whether it has changed from waking to sleep or identifying the several intermediate degrees between these two main points calculated in milliseconds. They measure that a certain configuration initiated the sleep process, automatically triggering the transmission for image insertion and total interaction, which begins in the visual cortex. That is, the person can dream immediately after sleeping. This stage will last as long as sleep is maintained or until it is purposefully interrupted to start the reverse process using nightmares.

Any content can be transmitted to be processed by the targets' minds. The content are usually movies that look like video games or interactive films that create situations in dreams in which the target is compelled to react according to what is most prudent at the time. It's similar to controlling a video game character while awake. The difference is that the character is your "self" within the dream. Moreover, you don't control the "game" using a joystick; you only use thoughts and mental commands that would play out the action in real life. However, this reaction occurs in dreams, since people generally don't know that they are in another reality while dreaming. They're unconscious and end up interacting and reacting according to what they would do in a similar situation when awake. The remote manipulation of the dream becomes a kind of "video game of dreams".

I know that all this seems a little absurd to you, but it's worth remembering that the MKTECH system is the most terrible weapon ever created by humans after the atomic bomb.

2.5.1 - Gamification of dreams

Gamification is a concept that consists of taking video games to places they've never been before or incorporating certain events and elements into the format of interactive games in which the player is rewarded or punished depending on their actions. One of the events that gamification has transformed was dreams. Following this uncommon premise, several tactics were developed to interact with the human mind in a state of unconsciousness. One of these stressful tactics is to force the target to play a game of any kind within artificial dreams. An act that in itself is degrading, but reveals something even more terrifying: the capture and exposure of the target's interactions in the face of situations created in this dream-like artificial reality, including their choices, dialogues, spontaneous reactions or deep decision-making.

This type of experiment exhausts the victim and causes massive stress in the dream that is reflected on the target's physiological reality. This ego [26]- thought-responses interaction to dream situations

---

[26] - The self, the "spirit", is the agent that controls our behavior and says who we are, composed of the personality and unique characteristics of each human being. The ego is similar to these agents;

has colossal commercial potential for the industry in general — entertainment, medicine, etc. — and infinite possibilities are opened up to be explored by operators regarding the theme, narrative and the content of dreams. To better understand how everything happens, I'm going to describe situations that were dreamed up by targets. These aren't hypothetical cases. They show how this dream gamification effectively works.

It all starts with the presentation of a setting, a dark atmosphere that resembles a hostile place. Then operators make the target believe that there is a gun in their hand. Soon after, a dangerous act occurs. Something comes out of the darkness and attacks them. As the danger arises, the brain will process the image and transfer it to its auxiliary systems in order to analyze the situation — which includes the limbic zones that act similarly in both waking and sleeping states. At that point, emotions related to fear and confrontation are triggered. The target — within the dream — in an egocentric and first-person perspective, shoots the gun while they're still trying to activate the electromechanical movements that would initiate the act to be performed in concrete reality. In other words, the body gesture that would recruit a series of muscles, place the torso in a shooting position, raise the arm, aim at the threat and pull the trigger. However, the system that controls the movements (frontal lobe) is deactivated in REM sleep. Then only the intention of the movement and its efferent copy — neural transduction connected to it without actually executing the planned movement — would remain. The neural intention is captured, amplified and converted into an **approximate return** action within the dream.

In the case of the motor system, when an area of the cortex sends a command to a muscle, that same area sends a copy of that command — efferent copy — to other sensory and motor structures that make the adjustments of perception and the postural adjustments necessary for the movement that is captured. D2K is able to convert this intention of movement into an EEG response — an electroencephalography, a method of electrical monitoring of brain waves. It works in parallel with the whole set of tools that monitor responses

---

however, it manages our reality when we're dreaming. It represents the modified consciousness that makes up the essence of the person in that state.

to the scene, processed by the adapted libraries and transformed into action within the game set up by the operators. In this way, the victim has the feeling of having shot with a firearm, of interacting and modifying something within the dream in conscious acts, but unconscious ones in sleep. In response to their action, the supposed enemies presented in the dream may or may not be shot. Regardless of the repercussion, visual and auditory feedback will return to the target's mind. In addition to that, the game, dream or interactive film has a sequence, which keeps the individual immersed in a fanciful electronic dream. The target is unaware that they're dreaming until they awaken from the dream-state. Therefore, all the action is experienced by the brain as subjective and "real".

We can explore another situation among several others that are created in studio and inserted on the dream. For instance, when the target is forced to push someone or something, e.g., an "actor" in the studio who runs towards the camera in a threatening manner. This camera is positioned so that it simulates a first-person perspective when sent to the dream. So, the brain assimilates the camera as being the ego's own vision, compatible with reality. To protect their body from the sudden threat, the ego (target) instinctively stretches out their arms to push away this frightening figure that is getting closer and closer. The natural intent of the defensive movement was captured at the very moment the act took place and returned to the studio where the interactive movie was being staged. Then, the actor in the studio was able to coordinate any scene by receiving feedback from the ego's reaction. During one of these sequences, a strange thing happened. In this case, the threat literally flew through the air after receiving the impact resulting from the collision with the ego, thus simulating an unusual push. A sense of super strength was felt by the dreamer right afterwards due to this unexpected return to the response of the act performed.

If it's perceived that the target felt an uncontrollable fear in response to the scene — that is, instead of pushing the threat the ego chose to stay in a defensive position or to stand still —, another actor's behavior can be triggered to increase the fear. Soon after, this action is sent back to the target's mind. There, the interpretation in the brain is generated and its output response is again captured and sent to the studio for adaptation and decision making. This creates an infinite cycle of interaction between the victim, the people and the AI playing with the victims' mind.

The act of pushing a person, running, deflecting a blow or preparing for a collision can be clearly synthesized in data and captured remotely, thus making it possible to adjust the theatrical performance in a new act to give continuity to the dream in the studio. It just depends on the degree of creativity of the operators and their intention, which can be waking up or stressing the target, discovering reactions, creating emotional states, and so on.

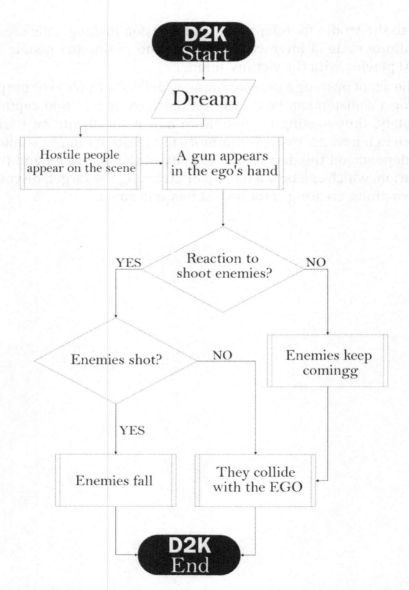

Figure 2.15 *The graph shows the flow of dreams produced by D2K based on the content reported above.*

Operators produce "live" footage with chroma key and CGI. They create the entire content of the dream, adapting to the target's responses to the presented scenario in a true theatrical performance. They always wear masks and costumes so that there is no possible recognition by the target. This gamification is very important for modern experiments, after all, it captures all attitudes generated in response to stimuli in the target that were created in the artificial dream!

When we're in REM sleep, the motor cortex creates organized electrical motor patterns that attempt to control the body, but without success. This helps the "self" within the dream to have the feeling that the body responds normally; that they're actually moving in the same way that they've always done since childhood using the acquired motor memory which is also carried on the unconscious ego. Thereby, the order to execute motor commands is maintained. However, the stimulus doesn't reach the muscles, it only feeds the fantasy in the dream world that is encouraged by the film sequence created in studio and is based on these captured reactions that don't activate the limbs. This deceives the brain and forces it to believe that the target is in the common reality (awake). Then, a tremendous mental effort and consequently great fatigue will occur as these manipulated dreams advance.

In the course of this chapter, we're going to see more techniques and dream situations and their consequences, always taking into consideration the concept of monitoring the reaction to the world of artificially created dreams.

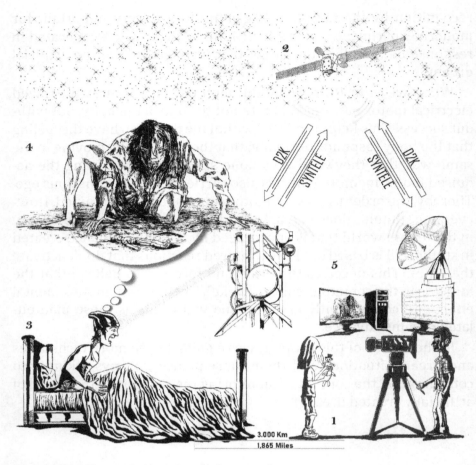

Figure 2.16 *Transmission scheme of remote dreams.*

1) The dream is recorded with professional cinema equipment, using various techniques similar to those applied to movies, on a remote base.

2) The content is sent to adjacent satellites and antennas in the correct modulation to be sent back to the victim's brain in real time.

3) Like an analog television, the target's brain receives the signal, demodulates the content and interprets the images also in real time.

4) In the studio, it's possible to create cinematic trickery that will cause severe nightmares and insert false long-term memories. The interpretation of the film — that was created in the studio — in dreams is extremely disturbing and can create more vivid nightmares than the target has ever witnessed before.

5) Return waves via SYNTELE (Synthetic Electronic Telepathy) create a response interaction flow for operators to "play" with the target's ego within the dream reality. Then they automate the feedback reactions, creating an interactive sequence between the "lives actions" (or improvised soap operas) and the answers given by the target in their own dreams. A real video game with the ego inside the dream.

2.5.2 - What is sleep and dream? How does D2K affect each stage?

Sleep is a transient and reversible state that alternates with the waking state. It's an active process involving multiple and complex physiological and behavioral mechanisms in various systems and regions of the central nervous system. We spend approximately a third of our lives sleeping and a quarter of that time actively dreaming — in the case of people who aren't connected to the MKTECH technology, of course. With regard to targets, the number of lived and prolonged dreams grows exponentially.

Modern understanding of sleep says that it consists of two alternating stages: REM (rapid eye movement) and NREM (non-rapid eye movement). Each of them has unique neural mechanisms and distinct electrophysiological and behavioral parameters. REM sleep starts in the second half of the night. This is when most long-lasting dreams occur. It's also called paradoxical sleep, as the brain wave activity is similar to that while awake despite being unconscious. However, there are also stages of transition in varying degrees between these two that are intensively explored by MKTECH, as we're going to see in the course of this chapter.

Dreams

Since ancient times, dreams have always been surrounded by mysteries and superstitions. Their meanings were interpreted based exclusively on subjectivity, such as windows to a new world, spiritual enlightenment, etc. It was only in the second half of the 20th century that the dream became the object of a broader scientific curiosity. The improved electronic instruments — used in modern sleep and dream research — discover, measure and record the tiny electrical potentials associated with all biological functioning. Now, scientists are able to distinguish certain variations in the bioelectric potentials that emanate from the dreaming brain and follow the psychological facts felt by the dreamer. However, they had no access to the content of the dreams until now.

The function of sleep isn't yet fully understood. It's only known that deep sleep is the moment when the brain performs a series of maintenance and synaptic cleaning processes. It's when the brain organizes memories, destroys the less important ones and deactivates some synaptic connections in groups of neurons and reinforces others that it deems more important. The automatic process of deciding what is important — and what is not — still needs further research for a complete understanding.

Some studies show that dreams result from a set of events: a) the selective activation of the **occipital lobe**, which is responsible for interpreting the visual world. In other words, it controls the processing of identifying different objects around us, differentiating and understanding different shades of color, etc.; b) the activation of the **parietal lobe** that integrates somesthetic stimuli for recognition, memory, spatial orientation and perception; c) the activation of the **limbic zones** responsible for emotions and social behaviors; d) the inactivation of the **frontal lobe** that influences the learned motor activity and the planning and organization of the expressed behavior; e) and the activation or participation of the **precentral gyrus** that lies on one side of the brain next to the anterior areas of the frontal lobe cortex — supplemental and pre-motor areas — and regulates specialized muscle activities on the opposite side of the body. The ponto-geniculo-occipital waves (**PGO**) with the Lateral Geniculate Body (**LGN**) activation would be responsible for visual images and limbic and paralimbic structures, and for emotions and memories.

The mechanism of dream formation isn't completely understood, but we know that the raw material of dreams is memorized information and active/passive and internal/external stimuli. There are several stages of sleep, and MKTECH operators are able to alter the natural functioning of the brain in each of them, thus resulting in unparalleled negative and positive experiences and sensations for the Targeted Individual.

The meaning of dreams, how their content is generated, their ability to influence every organism and why they exist are still debatable. Some studies indicate that the act of dreaming is associated with learning and memory, which includes "virtually" facing traumatic experiences and emotional adjustment. It improves the state of mind, memory and other cognitive functions through the recovery of certain neurochemicals that are depleted during the course of mental activity in the awake state. Dreams are also thought to be used to test brain circuits linked to behavior and primary cognition, and to keep mind and brain in step.

Other theories suggest that the act of dreaming exists to prepare the dreamer for future experiences, some of them connected to emotions and memory. In more recent studies, it has been found that dreams are also linked to the mesolimbic-mesocortical systems known as the brain's reward system. This fundamental system is responsible for the motivation and interest in everyday things and in life. That is the reason why this gamification of dreams holds the ego strongly in the synthetic narrative and affects the whole organism in a negative way each night. These dreams explore our system of expectations related to reward and punishment, as it always presents themes that causes maximum pleasure or extreme suffering.

Now I'm going to explain how the process that triggers the shift from the awake to the sleep state occurs.

NREM sleep is divided into stages 1, 2, 3 and 4, and it occurs predominantly in the first hour of the night. After lying in bed for a few minutes in a quiet, dark room, you're probably drowsy. Your subjective feeling of drowsiness is registered by a variation in brain waves that highlight the following explanation: eyes closed, occipital alpha waves, waves from **8-13 Hz.**

Stage 1 — Low-voltage waves of **4-7 Hz. Theta waves** is a very light sleep state, described by several people as "drowsiness" or "tendency to sleep". You drift in and out of sleep, and wake up easily. The sudden "falling" sensations that wake an individual up in fright occur at this stage. There is a decrease in muscle tone. At this point, the victim's brain enters an electrical pattern susceptible to replacement of normal thoughts through the transmission of electronic dreams, which initiates the reception of the images transmitted by operators. It'd be like you had those old analog (cathode-ray tube) TVs and slowly tune in to any channel. Little by little the images are formed in the midst of a lot of static and noise until you're able to tune in the desired channel and have a well-defined image of the transmitted content.

This is what happens with the victim's brain when they embark from an awake state — in which they control their actions — to a state in which the contents of thought are no longer controlled by them, based on complex chains of influence in internal and external sensory information. Your complete "self" is gradually disconnecting. From an intermediate point to the first stage, in the midst of this transition, invasive transmissions are slowly received, interpreted and sent so that the mechanics of sleep and dreams naturally absorb the content while using the target's brain as a data processor. Some say that when we dream, the mind awakens and the body sleeps.

At this stage, operators usually send threatening messages. They use images with grotesque manipulations to prepare the target for the new type of horror adventure they're about to embark on. The microwave voice is in charge of the auditory part, which has a strong influence at this stage. This technology degrades the sleep quality and is responsible for waking the victim up with a scream that can echo throughout the brain, since the process of inserting data directly into the mind of the unconscious target has already begun. It's precisely at this moment that the phenomenon many victims report as overwhelming is experienced. **"Tunguska sound"** [27], a bang so intense within the auditory cortex that it is capable of waking the individual right away. In certain circumstances, it can cause permanent neurological damage. More information on this phenomenon linked to V2K/SYNTELE stress and the different degrees of transition from conscious to unconscious can be found later on this chapter.

All changes are carefully monitored. They contemplate the absolute knowledge of the transitions that occur in small subtle changes

---

[27] - The Tunguska event was a meteor fall that occurred in a region of Siberia, in Russia, close to the Podkamennaya Tunguska river on June 30, 1908. The fall caused a huge explosion and devastated an area of thousands of square miles. The event generated one of the most intense sounds ever recorded in history.

in the set of remote EEG configurations and other diverse analyzes that indicate each phase and subphase related to the target's sleep with great accuracy, which compares a data set that configures the current state with the same sets of parameters already known to the AI in its system database. It thus automates certain attacks, diversified at each stage without the need for human interference.

Stage 2 — As soon as the person falls into a deep sleep, there are large amplitude biphasic waves with acute negative phase and slower positive phase. The so-called K complexes that are characterized by the presence of spindles — the "sleep spindles" — in the sigma frequency range of **12-14 Hz**. The body cools down and the muscles begin to relax. There will also be brief bursts of brain activity, usually associated with muscle spasms. It becomes more difficult to awaken the person. It's understood that the individual has already slept, occupying 45% to 55% of the 8-hour sleep.

It is at stage 2 — where the consciousness is slowly making room for the subconscious — that dreams with images that lead to sleep, the "hypnagogic" ones, are usually experienced. If you were awake at this point, you might as well describe vivid and grotesque images. Most people have had hypnagogic dreams. They're almost hallucinations of sleep, capable of causing strong emotions and generating long-term memories modulated by this emotional charge. In fact, this charge is overexploited by MKTECH operators as a technique for recording long-term memories in a state of unconsciousness. According to popular belief, this nightmare is famous because it has a mystical quality. In some parts of Brazil, the appearance of the Brazilian folklore character called ***Pisadeira*** [28] is reported and is still mistaken for abductions, encounters with extraterrestrials or contact with gods.

Operators and synthetic electronic dream technology don't decide what kind of dream a person will have — hypnagogic, REM or

---

[28] - The character is usually described as a very thin woman with long, dry fingers and huge, dirty and yellow nails. She has short legs, shaggy hair, a huge nose with many hairs growing out, like a hawk. The eyes are fiery red, evil and wide. The chin is turned upwards and the mouth is always wide open, with greenish and exposed teeth. She never laughs; she guffaws. A shrill, horrifying laugh. She lives on the rooftops, always on the prowl. When a person dines and goes to sleep on a full stomach, lying on their backs, *Pisadeira* goes into action. She comes down from her hiding place, then sits or steps heavily on the victim's chest who enters a lethargic state. The victim is aware of what is happening around them, however, they're helpless and incapable of any reaction.

none. They don't have the ability to influence the mechanics of the sleep stages nor are they able to modify primitive characteristics, the "engine" that is embedded in the "natural neural algorithm" of this step by step. They just take advantage of the situation and stimulate a certain stage based on sleep deprivation and recurrent stress. The stages, however, remain. Although the reason for their existence is unknown, hypnagogic dreams continue with their natural configurations and functionality for the mind. So, when the configuration mapped by the technology — and based on a set of received data — indicates that a dream with the approximate hypnagogic characteristics will be displayed, operators replace the target's dreams with images transmitted straight from their remote base in an appropriate format to be presented within the dream. They then await confirmation that this complex combination of factors capable of interpreting the image with these unique *hypnagogic* attributes has in fact been inserted. At this point, some settings are systematically checked for the system to send the dreams that best fit the bizarre natural characteristic and state of mind of the target at the moment.

At this stage, operators are ready to repeatedly send the transmission content that will replace the sporadic *hypnagogic* images. When the brain prepares itself to display the naturally created images, the transmission will replace the natural dream with artificial images created electronically from a remote base. As soon as they notice that the victim is processing the first images with the invasive content of the dream, the rest of the electronic film is released. It'd be the equivalent of a device successfully connecting to a remote server in terms of computing. The operators then display the first images in a loop until they obtain confirmation that the correct natural configuration of the brain — capable of capturing, interacting at the exact frequency and demodulating the content of the artificial dream — has been achieved. At this point, the victim cannot distinguish whether the dream is natural or artificial. So, as the brain is connected to the transmissions, the loop is unlocked and the film continues. It becomes the world, characters, scenarios, sounds, colors, the total lack of laws of nature (such as gravity, for example), the story, the beginning, the middle and the end of their reality.

In the next step, the victim's brain is under the total control of the operators with respect to the content displayed in the mind and the psychophysiological responses. In the case of hypnagogic dreams that have a limited visual component and a macabre atmosphere, the brain displays and interprets these images only in gray and black. It's very different from a REM dream, which tends to be longer with more vivid colors and a more complex psychological immersion. The characters that make up these dreams range from demonic creatures, terrifying beings, zombies and well-known characters from literature, movies, myths or games.

Neither that sleepiness after lunch, the afternoon that the person eventually takes off, nor the famous "nap", escape. As soon as the person falls asleep, D2K masters the content of dreams in a matter of minutes. The characteristics of these dreams are generally similar to hypnagogic images. In some cases, a brief 30-minute rest is already enough to generate relatively long dreams in the distorted perception of time in the particular fantasy world, in relation to the amount of time spent in ordinary reality.

Another point that differentiates it from the REM dream is the highly-charged emotional nature. Taking advantage of this favorable opportunity, operators exhibit films and images that will trigger very intense feelings and emotions with direct negative consequences for the body. It can cause tachycardia, excessive sweating, the feeling of having a pounding head and other physiological symptoms, such as increased blood pressure, headaches, emotional and spatial disorientation. The first times the victim is subjected to this type of weapon, the emotions provoked by the dreams are so intense that this experience will probably never be forgotten by the human subject.

Taking advantage of the nature of hypnagogic dreams, operators make the victim feel extreme fear with scenes of death, accidents, wars, monsters, voracious animals that appear to come out of dark places and chase the target. It always culminates in violence and makes them wake up extremely frightened and disoriented. Operators even simulate a recurring dream of someone chasing the ego to their childhood home — a house that was incorporated into the film and captured from their visual memory. The target arrives at the door of that house and ends up trapped. In the subsequent act, the pursuer pulls out a gun and shoots. This produces primitive emotions, such as the feeling of imminent death, intense fear and acute stress just when the brain is supposed to rest and invigorate as it prepares the body for the next day ahead.

The primary emotions generated are varied — from the feeling of being in front of a very powerful being, the sensation of falling or drowning to appearances of mythical beasts and mystical creatures. They use known natural experiences and apply them to electronic artificial dreams, however, with more expressive and accentuated results. The feeling of revolt, sadness and helplessness upon awakening and after a few minutes realizing that your dreams were manipulated in such a dark way is also felt. The target is usually awakened after a hypnagogic dream. A memory added to an overwhelming emotion is inserted and will later be treated by the brain as a long-term memory, but without going through the normal process of this type of storage, as it would normally happen when awake. At this stage, the victim doesn't interact with the dream. Intense emotional changes unfold, just like in a movie. Actually, it *is* a film edited and transmitted by operators.

Here are some consequences of having an artificial electronic dream and being awake at this stage of sleep:

* Headaches;
* Increased heart rate;
* Mental confusion, long-term disorientation;
* Sudoresis;
* Helplessness in the face of the situation;
* Long-term memories that will remain dormant in the target's mind for their entire life.

Stage 3 — It's the first phase of deep sleep: slow, low-frequency, high-amplitude waves. This stage is characterized when at least 20% of a period is occupied by delta waves (**1-2 Hz**) until the EEG aspect is completely controlled. Muscle tone decreases progressively. If you're awake during this stage, you may feel weak and disoriented for several minutes before regaining full consciousness of your surroundings and actions. The hypnotic stage of obtaining information occurs during this period. At this state, transmitted dreams serve to extract information from the target.

Do you remember when I said operators record familiar voices that were extracted from the day-to-day social interactions in the chapter on EMRa? At this stage, these voices are used to reinforce the target's sense of security so that they talk openly with their attackers. In the *hypnotic* stage, the victim becomes highly suggestive. You may have already witnessed this stage in practice. The person next to you falls sleep and sometime later they start talking in their sleep. That way, you can interact with the individual in a long conversation if you have access to this technology. You'll able to connect the vocalized thought using SYNTELE and check all internal answers to the questions asked.

From these dreams, operators obtain information about important secrets and intimacies of the target using SYNTELE and previously recorded images of everyday life. The V2K — sounds in the auditory cortex — creates a more real experience, as it provides the victim with the impression of the auditory similarity that one would have in concrete reality, maintaining the dialogue with subvocalized thoughts captured by EMR. The targets don't know if they're awake or asleep at that moment, so debates about their intimacy and the intimacy of the people in their social circle are usually posed, such as customs, prejudices, preferences, moral character, among other diverse traits that make up the essence of their consciousness. In this way, all reactions to questions raised on a cognitive level with the intellect partially functioning are exposed, leading to automatic and true responses that could be different if they were consciously analyzed and answered verbally.

There is little interaction with the visual part of the dream at this stage, however, the target's mind is guided by an operator's voice via V2K that works with the imaginary and memories. A patient having an open and straightforward conversation with their psychiatrist is what comes closest to this interaction in ordinary reality. In this case, however, the victim is obliged to submit to this type of awkward conversation. Operators directing this interaction are enemies who want to destroy the target's life and have fun during the process while collecting as much data as possible for modern experiments, conducting tests and playing with private memories. It's a complete violation of the democratic freedoms and privacy rights obtained with great difficulty and infamously perpetrated by the bandits behind the technology.

Stage 4 — When the proportion of delta EEG activity exceeds 50% of a period, the Third Stage becomes the Fourth Stage. This is the "deepest" stage of sleep, in which **delta** waves predominate. It's the most difficult stage to wake the person up. During this period, the electroencephalographic characteristics present **theta waves of 2-7 Hz** and, at some point, present sawtooth waves coming from the frontal region of the vertex, which initiates the REM sleep stage.

After a period of REM sleep that lasts perhaps five to fifteen minutes, it's common for the person to go through the previous cycle again, dreaming vividly three or four times during the rest of the night with two important changes. In each successive cycle, decreasing amounts of slow-wave EEG activity appear — Stage 3 and Stage 4 or delta sleep. Later at night, perhaps after the second or third period of REM sleep, no delta sleep occurs: only REM and the second stage of NREM sleep take place.

During REM dreams, the mind control technology — more specifically D2K — starts its deepest process. In this period, the abstract and extremely psychological madness uses the brain's dream platform to navigate the cortical modules without the "self" filter, the awareness of the awakened state itself, which has mastery over rational thoughts and is able to decide about various aspects of reality.

The greatest cognitive violation known to humankind begins when the "self" gives rise to the ego.

Function of dreams in REM sleep

Dreams have always been main discussion topics since the beginning of the humankind. Several psychological and philosophical theories and approaches are used to understand them. They're known to be important for the development of a child's brain, for example, as it provides an internal source of intense stimulation that facilitates the maturation of the nervous system, thus allowing the testing and application of genetically programmed behaviors without the consequences that could be fatal in real life.

Directing their dramas in dreams, with complete mastery over the content, makes the Targeted Individual a *testing machine*, and all kinds of topics can be transmitted to them. Themes normally vary, ranging from subjects based on their private life, as it explores their human traits, to totally disconnected themes that mix live theater, cinema, games or montages that create fanciful content to the extreme. However, there's always some secret mysterious parameter, or scenes to be tested in the experiments conducted in dreams.

It's an extremely personal and unique process, adapted to each individual and their particularities, experience, reality and life experience. For it is known that this remote manipulation of dreams, although exceptionally destructive, is only one of the modern torture techniques using MKTECH. The D2K serves as a supporting tool to find out more about the target's life and to resolve any doubts that may exist. After a full day of intense, almost unbearable SYNTELE (Synthetic Electronic Telepathy) bombing — bearing in mind that the individual is already exhausted from the silent, external struggle that is an absolute hell inside the mind —, the target is still deprived of rest, as it's practically impossible to sleep with the shouting via V2K within the auditory cortex. If by any chance they manage to overcome this barrier, the target will later suffer in each stage of sleep.

2.5.3 - Dreams created by D2K in REM sleep

In this part of the book, I'm going to report some real experiments that were — and still are — performed with the victims using synthetic dreams, and their consequences and abnormal reactions that gradually modify the targets' perception of the reality around them. The cognitive violence of this act results in serious repercussions in all aspects of life. For the dreamer who had their brain attacked and used as an image and sound interpretation experiment, this becomes a remarkable event. The artificial modification of dreams content in the REM phase occurs when they experience highly complex and long-lasting dreams with the capacity to make the brain experience unique moments that would never happen while they were awake. They insert the target in these narratives and give them roles that vary between the lead actor and a passive, observer character.

As stated at the beginning of the chapter, a person's personality when they fall asleep remains active within the dream. Traits — such as common sense, intellect, intelligence and moral values acquired during the course of life — are used to judge the situation and to react to the images and sounds sent by MKTECH. Thus, the target is forced to expose their privacy in the face of a situation, even involuntarily. Don't forget that they are sleeping and don't even notice. The target embarks on an epic saga of dreams and transcendent experiences that have active and destructive responses in each scene that unfolds. So, operators get to know the target's interior, their limitations and qualities through direct experience.

The brain needs parameters to react to this artificial stimulus within dreams, as it is unable to display the scenes in an impartial way and without affecting the rest of the organism. Due to this brain feature, it's possible to create credible stories with real and imaginary visual memories, using characters known to the target: friends, relatives and spouses, for example. It becomes then possible to conduct a soap opera with deep interaction between these characters, and to capture an infinity of private stories in which its development suppressed feelings, resentments, joys or deep buried secrets that are part of one's essence. Keeping the above in mind, it's easy to simulate a girlfriend inside the dream and create a situation so that all secrets of the past are instantly exposed. Just reproduce scenes that have characteristics similar to those of the past and recreate them in studio using the target's memory, the operators' theater and montages and techniques created on the computer. Consequently, they delve deeper into the study of the target's private identity, strengthening everything that is destructive, such as fear, insecurity, frustration, loneliness, emptiness and acute stress. All of these emotions can be used, but harmful ones are the most sought-after by operators. Each striking image is associated with strong emotions, and it strengthens highly-invasive and vivid long-term memories. They can even incorporate tactile sensations from the moment the victim dreams.

A more common example of this situation would be to sleep and end up in an unfavorable position. Your limb would be compressed, stifling blood flow to the area. This sensation can be detected and incorporated into the dream. For example: a leg goes numb and this somesthetic stimulus starts to affect sleep (and the dream) when trying to wake the individual up to modify their posture and reestablish the correct blood flow. This "order" is then captured by both the central nervous system and the operators and, automatically, images of a mutilated or deformed leg will be inserted into the dream. This confuses the brain, increases the immersion and its subsequent consequences when the target wakes up with terrible memories. After all, the tingling pain is linked to an extremely graphic and violent image: the person looking at their mutilated or crooked limb.

Operators often "edit" dreams using cameras that includes gyroscopes and accelerometers (e.g., a GoPro). They simulate first-person perspective and create dreams interpreted as highly realistic, with sharper images than in any other ordinary dream. In this way, they can synchronize the interpretation of the tactile transduction of any nature from the real world with images within the dream, contextualizing these sensations in the story that is being filmed in real time and transmitted to the target. Bear in mind that dreams can have an impact on the dreamer's brain as great as that caused by a corresponding concrete sensation. The sensations caused by the dream can seem very real or even "more than real" for the duration of the dream and a few minutes after waking up.

Manipulated dreams may contain content that profoundly affects the target while awake. They can impair the way the individual behaves for days, in addition to their whole set of daily activities, especially work and social relationships. They can also affect the learning process, since access to memory and attention will be focused on the new memories forcibly acquired via D2K of a reality that only exists for the dreamer. For example, scenes of loved ones being slaughtered or murdered — the creation of an atmosphere of terror and loneliness in which you lead a life of utter failure or extreme suffering. Any virtual reality can be created. Operators have deep prior knowledge of the target's psyche as well as their fears, or they acquire such knowledge in the course of these artificial experiences.

Let's take a target who is affected by a recent loss as an example. This sad event will be explored in dreams, and will leave deep scars. The traumatic memory will be relived several times and will worsen their physiological state — that was reached during the dream — in the concrete reality. They will then become depressed and their day-to-day willpower will be destroyed. This enhances a bad general condition if the victim is going through an adverse moment in their life. Moreover, the attacks happen on a continuous basis for a few days until the effect wears off. After that, they use all thoughts captured while awake — when the target reflected on their suffering and the theme of dreams. This series of emotionally disturbed particular thoughts is also used as a new aspect of the same subject to be explored in order to make possible the recurrence of the peak of an initial painful and depressed state, which was achieved using the main theme and which over time naturally weakened. It's indeed an extreme violence committed against the thinking of others.

Another example is the famous dream of falling from a high place. Most people have experienced this type of episode in their dreams and know how disturbing it is to wake up after a fall so realistic that it generates mechanical reflexes in you — usually a kick that comes with a reversal of sleep rhythm. This experience usually occurs only a few times in the life of a normal person. For a MKTECH target, however, this is very common. Waking up provokes traumas in the world of dreams and unique psychological questions with profound reflections on the concept of reality. This dream mixes cinematic effects and the use of special camera angles. They normally film the fall scene from the perspective of first-or third-person view simulating the ego, in addition to using pre-existing models and images of any fall from movies or the internet. It may be filmed in a unique way never experienced, so it's assimilated as a great experience for the human brain.

I particularly never heard any news about someone who had natural dreams that use special resources with the generated scenes, for example, using virtual glasses or a 3D recording with high-definition images from HD to 4k. Yes, reader, operators conduct experiments using footage with this type of device to raise the level of immersion in the dream. What happens when the brain receives this type of footage is simple: it goes "crazy". The level of immersion in such dreams with this *neuro-cinematographic* novelty — that completely modifies spatial perception — causes extremely relevant disturbances in the brain, especially when it comes to this new reality. Dreams processed by the mind, with virtual or 3D characteristics, are assimilated in a peculiar way, as they record vivid long-term memories upon waking up together with complex harmful physiological effects.

Generally, dreams that violate the laws of physics — flying, being weightless, being stuck in the water with no equipment and not drowning or swimming in incandescent lava — result in instinctive actions to predict a pre-existing memory. They trigger defenses, such as expecting to be burned, running out of air, or dodging something that comes your way, which reflects on the organism as an action. They can recruit instinctive actions of this extremely realistic nature, in which a scene interpreted by the brain as threatening is displayed by activating the limbic systems. These actions — reflected in the nervous system and in the interaction of the ego within the dream — add several valid data for operators.

Another dream that wakes the target abruptly is created as follows: there is a calm and tranquil atmosphere with relaxing landscapes, feelings of well-being, peace and calm that will unburden the ego accompanied by rare positive emotions, which confirms the possibility for a beneficial use of this tool. Suddenly, a quadrupedal "animal" — zebra, horse, mule, Minotaur, Satyr or a Centaur, for example — bursts in a flash and violently kicks something in the dream that the brain interprets as an imminent aggression to the "self". The mental security system is activated; the instinctive reflex to dodge the danger and the act of waking up is triggered under the same later consequences. The target wakes up with several physiological problems resulting from stress: sweat, tachycardia, rapid breathing, among others.

Keep in mind that the dream "self" (or ego) has a set of basic qualities and traits that remain active for the mental system to function in another reality. For instance, when you remember that you are *you* within the dream. Otherwise, we wouldn't perceive the dream in our minds and we would have nothing to rely on in terms of rules and primordial principles that process and compare, and we wouldn't understand a certain event in this manipulated reality, even the most basic one, such as identifying the image of an ordinary object acquired in concrete reality and understanding the meaning of it through its mental representation. This encompasses a set of characteristics that synthesize the information and immediately tell the ego that that character is dangerous and should be avoided; that a car is a car, an animal is an animal, and that the person in front of you is a member of your family. Only then you can respond to the challenges that arise at all times for the brain to analyze.

This primitive collection has several human characteristics that influence everything, especially the responses given through imposing situations. In these situations, answering them depends only on the essence and personality intrinsic to the human who dreams — and not just the mind lacking the rational control that manipulates afferent signals flowing from the common reality. This pure, immediate, naïve and sincere response from the ego is what reveals absolutely any secret of the target's life, by interacting in an atomic way with fragments of the synthetic electronic world.

Another very recurrent and traumatic dream begins by presenting a scene of danger from which the target is running, fleeing and being taken along several paths, which causes a false sensation of consciously directing one's dreams. Then, suddenly, a ladder with its elevation highlighted by 3D footage is presented and they fall, even if the target's desire in the dream is to stop and descend normally. This intention is captured by MKTECH, however, the image execution against the target's conscious and voluntary will within the dream remains. The ego is driven to speed up and hit or trip over this obstacle, making them wake up abruptly, feeling sick and terribly shaken by the experience. In this type of dream, the consequences of opposing the dreamer's will are explored. Can you see the subtle details that these artificial dreams are capable of producing? In this particular case, incoherent stimulation was explored, a known technique of conditioned dissonance designed to turn a person's personality inside out. Thus, the real tests and experiments in a higher sphere that we're going to see in the following chapters begin to become clear to the target of psychotronic torture.

Other type of dream that can affect the target in several different ways and test the effectiveness of some theorized and implemented methods include sudden movements, 360-degree rotation, flying, running and falling. Some dreams achieve this goal by using footage taken by drones with their spatial movements. The initial flight that makes a panorama with the insertion and mixing of other elements causes the same *sensation* of flight on the target due to the mechanics based on mobility similar to the movements of a helicopter. You can rotate 360º around its axis, as well as perform movements horizontally with the perspective fixed at a point on the horizon, vertical movements and variations between these axes. By moving the cameras, infinite possibilities are created in terms of impressing the brain. An exact correspondence is also created between the direction of the eye movements of the person who is dreaming and the direction in which the person is looking in the dream through the drone's cameras. The experience of flying is assimilated with such intensity that is rarely observed, even in natural dreams and nightmares.

A technique similar to the previous one that confuses the brain is to use dolls or models to compose the dream scenario and shoot to scale with a camera attached. This camera simulates first-person perspective, flies over places, houses, castles and other structures, in aerial shots similar to resources used in cinema. It works very well, since the dreamer really has the feeling of passing by these structures while flying, swimming or inside vehicles. However, at a certain point, the dreamer acquires the ability to perceive that the dream is artificial due to the reduced scale of the apparatus used. It takes some "dreams" for the target to notice though. Over the years, neural adaptation takes place and the target is able to distinguish the objects used — in addition to clearly recognizing toys, places, miniatures, basically everything that was used to create the dream — upon waking up.

The scene unfolds within the dream. It begins with the feeling of being in a fake world. Bandits are spotted on the horizon and are getting closer and closer. The sky is dark; the scenery, intimidating.

There is an atmosphere of destruction and chaos in the air; the feeling of dread is getting worse. Then a loud, high-pitched voice screams: "They're coming. Run!". Your legs don't obey the command to move. The bandits arrive and start to attack: blood, stabs, punches. You enter an artificially created world that is menacing and unpleasant with scenes carefully elaborated to cause the greatest terror and fear possible. This is another interactive film made in studio that depicts the content of a human guinea pig's daily dreams. The emotional excitement provoked by dreams like these spills over into reality once more. This type of disturbing nightmare that occurs on a daily basis gradually affects the target's quality of life as a whole.

It starts with a calm scene. Nothing is happening. Suddenly, the ego finds itself passing by a certain unidentifiable person who wears a mask in the dream. There is a sudden movement, apparently a blow to the head, a violent headbutt, which wakes up the victim. It's then clear how the scene was recorded in the studio: an operator with a first-person camera passes by a certain place — which simulates the "self" point of view in the dream. The blow seems to come on the dreamer's face as if it were in real life; an actor headbutting towards the camera. As soon as the target wakes up, they feel something strange around the region where the blow was struck. However, they don't actually feel the pain, nor the subsequent consequences of the attack, such as swelling — a natural reaction of the affected area if the same act occurred in real life. Even so, this raised doubts as to the true intention of dreams systematically reproduced in this format, which makes evident the search for effects similar to the reality of an impact.

Experiments disguised as dreams attempt to achieve the sensation of pain without actually stimulating it. They're trying to perfect this weapon to the point that it may be possible to eliminate a target based on an excruciating pain in dreams, provoking this sensation as if the act had actually occurred in real life. Or perhaps the intention is to cause physical neurological damage — the result of brain interaction with transmissions — and, later, the sensation of pain would be triggered by the natural injury of the event. In both cases, the pain would be felt. I don't know if it's ever possible to achieve this result, but the intention of the experiments is clear. Acts like this are constantly presented in dreams amid several other tests related to pain. The attempt to activate the sensation of pain only through the reality inserted in the brain — without the actual physical contact that causes damage to the body's structure — is one of the scopes of modern experiments disguised as torture. By the way, this outcome will be the "Holy Grail" of remote torture if it is achieved. Those responsible for creating the films for the dream will be infuriated, since they have a perversion in cowardly torture and in the suffering of others, as we're going to discuss in chapter 5, Volume 2.

Another dream that occurs frequently happens as follows: the brain believes that the ego (target) is losing all of its teeth and is chewing them simultaneously. The sensation of loss is of sensory quality similar to what would happen in reality. The sound design that follows the event is emitted via V2K, simulating the grinding of teeth, the friction between them, and the *"crunch crunch"* from chewing is computed and recorded in the target's memory. The emotions of living without teeth are also experienced. The only sensation that isn't evoked is that of pain, which in reality would be unbearable. The pain associated with plucking teeth or teeth mysteriously falling (in dreams) doesn't happen. The search for emulated activation of pain is a maxim in dreams. In addition, there is a similarity in the patterns when the dream uses simulation of fights, wars, flight, falls or of being run over. The result is always the same, which culminates in a traumatic event.

Once, an extremely real and compelling dream ended up revealing another technique widely explored by operators: during a helicopter trip, the vehicle was placidly hovering over green valleys and paradisiacal landscapes. There was a sense of peace rarely felt throughout years of torture in which the target was forced to think and dream whatever the operators wanted. After a few minutes, the helicopter calmly descends. Suddenly, the target is at sea and sees a gigantic wave on the horizon that seems like a rogue wave. The natural fear that a wave of that magnitude would cause in most people immediately comes up. In this case, the target is used to the sea and knows how to swim in concrete reality. However, the most incredible and unprecedented act occurs after the impact: the target's immediate reaction is to close their eyes and dive as deep as possible.

As soon as the rogue wave hits them, they feel a strong and realistic drag, identical to what would happen if it happened in concrete reality.

This dream tells us that as soon as the wave broke, an actor was filmed in the water — probably a swimming pool — with a camera strapped to their head in order to emulate the first-person perspective of the ego; only arms and legs were visible. Then, it was only necessary to move their hands as if they were struggling in the water while trying to get out of that whirlpool; their feet followed the same movement, and then they were whirling and somersaulting with muffled and desperate screams. In this way, an atmosphere of singular realism is recreated. CGI is also used to give the final touch by adjusting the appropriate lighting, creating "foams" in the water and some details that cause the perfect visual impression of being dragged under the water, which perfectly reproduces the sensations of this act.

The most disturbing thing about all this is to notice how a virtual component is able to simulate the dynamics of the real event in the human mind and still compel the body to recruit defense centers in response to the threat by "triggering" a real danger alert. This causes the target to activate the posterior parietal cortex — place where movements start to be planned — in an attempt to defend themselves when thinking about the movement's motor action, the reflexes, as they imagine the defensive position of arms and legs within the dynamics of the event. In this process, the nervous system is affected, reflecting in physical symptoms of nausea and dizziness as the "body" reacts to being carried away by the water amid dizzying somersaults and twirls.

Anyone who has experienced this at sea — even if there were medium waves instead of rogue waves — knows how it feels. The sensation created by D2K can be so vivid that the body and the mind may believe that this is a real event. This always reminds me of experiments conducted for different purposes, together with the attempt to activate the pain just by the expectation of the episode conducted in a visual and auditory way. An unbelievable occurrence that only the dreamer who experienced the synthetic event understands the impact on physiology and the use of this weapon as a center of neural entertainment for commercial purposes. A kind of dreamlike Matrix [29] emerges as a promise — this is the only definition I can give at the moment about the aforementioned events and experiences.

Waking up the target by synchronizing with dreams

Given the observation of several dreams with different contents that result in the same final effect, empirical evidence leads to the belief that there is an ideal point in the convergence of certain physiological, electrical and psychic configurations. That is, a set of attributes capable of opening a "temporary window" that indicates to operators, by means of inference algorithms, the best moment to awaken the victim. This unique moment can potentiate the impact of the harmful effects of a traumatic awakening, and causes a remarkable experience at maximum intensity associated with emotions and latent memories of great availability when awake.

The improvement in the calibration of these parameters, until a common point is reached, is part of the fragmented dream research in progress. For this, it's imperative that the individual is unaware of the existence of this weapon. Thus, they use random people all over the world and subject them to despicable tests like these, violating their personal space and private life. Under other conditions, the results of these experiments could never be achieved if conducted in this format in a controlled environment interested in all aspects of dreaming. Surprise, ignorance and the dream "glimpse" by the target are key components for a satisfactory result.

It was also possible to verify that the originality of each image and situation is what causes extreme emotions. The first modified dreams are undoubtedly the most impressive ones and those that most create residual memory for the waking state. The target may not even know in what reality that situation happened.

---

[29] - The movie describes a future in which reality, as perceived by most humans, is actually a simulated reality called "the Matrix". This reality was created by intelligent machines to subdue the human population while using their brains for distributed data processing and their bodies, which are asleep in suspension chambers, as an energy source. An original story written by Lilly and Lana Wachowski, which was later modified by the film studio.

Note: as you read the book, some of you may be wondering why I usually use the word *dream* and not *nightmare* since natural nightmares are similar to those portrayed here as dreams. The similarities consist of scenes and situations filled with dread — the individual remembers everything in detail —, and include themes evolving from threats to survival, security and self-esteem, a high degree of autonomic discharge. It'd be a mixture of dreams and nightmares, but we're going to use the term *dreams* here. The nightmare will be present if the operator wishes, otherwise, a peaceful and immersive dream may happen. In addition, nightmare and dream can become subjective concepts over time.

The subtlety and the grotesque mingle with the silent torture in the target's mind perpetrated by operators in daily experiments. Did you notice the amount of information that is captured in each dream? It's impressive! From decision-making to sexual stimulation, or just sleep deprivation through any content that interferes with normal dream mechanics. This is how the victim becomes a false oneironaut. They can also comprehend the same phenomenon through multiple interpretations and different simultaneous perspectives — both the perspective of the dreamer and of those who produce the dream —, thus improving at each event the details of the final product, the film.

When we're awake, aware of the reality around us, and suddenly we fall asleep and don't even realize it, we tend to ignore the transition time as our conscious mind partially shuts down and we move from one reality to another. We tend to believe that we didn't leave concrete reality behind and that both events happened immediately. This makes us feel as if we're awake, so we enter the world of dreams and encounter a setting similar to the reality. For example, the target (ego) that woke up in the dream and came across the landscape of the neighborhood from their childhood — the image can be acquired in several ways, from capturing the target's visual memory, filming on the spot to taking photos from the internet — in which several striking interactions took place, always emphasizing the scenario. The ego better absorbs the situation when something like this occurs. It's like reliving their own childhood intensely, or at least until the target awakens and deep, antagonistic emotional conflicts take place. They feel a mixture of enthusiasm, sadness, satisfaction and dissatisfaction as the brain regains full consciousness and leaves the dream configuration in deep REM sleep.

Other widely explored dreams involve the use of spaceships and encounters with fictional beings or extraterrestrials. They use familiar everyday scenarios from real life mixed with spaceships coming from the sky. They also use CGI to merge scenes from films with the memory taken directly from the target's mind to compose the dreamlike landscape, thus creating a unique, emotional and intense experience that the target won't forget for a long time.

The individual in a first-person perspective — which is commonly used — observes through the window of their old childhood building and sees the place where they were raised, while experiencing timeless memories. As they watch the sky, lights start to flash. An object positioned in the distance with a precise and detailed depth emerges and goes towards the "person" within the dream. The camera follows the movement, then there are several UFOs and alien spaceships created using CGI or miniatures that go through the landscape where the target is. At that moment, the emotion felt by the dreamer is as intense as if it were completely real. In fact, it's not uncommon for dreams manipulated in this way to seem more real than reality itself, because that kind of emotion has never been felt or evoked before. After all, we never experience encounters with spaceships or extraterrestrials in concrete reality.

Therefore, the memory of a dream of this nature — that mixes scenes from the past actually experienced by the victim with special effects — is "recorded" in the mind as an experience so real that it will probably never be unavailable and will remain easily accessible while awake. It'll become an explicit automatic long-term memory, thus crossing the barrier of dreams. This experience, for those who were unable to see what is happening behind the scenes, is an experiment with a very powerful commercial and military bias as we're going to see in the following pages.

Dreams about alien abductions and classic clinical tests, deeply rooted in popular culture, are also created by operators. They can be recurrent and temporally united by fragmented events similar to a TV series in which the plot unfolds with each episode. These effects are purposeful to once again hide the truth by using popular beliefs. The creation of the illusion is able to make the target swear that they had real encounters with aliens. However, these are just perverse manipulations of experiments remotely conducted on the brain. The degree of conscious perception added to the sharpness of the images processed by the visual cortex generates this type of confusion.

The strategy of creating facts in dreams is similar to a soap opera: a series of episodes filmed with real people in the studio and sent to the target's mind in order to be demodulated in stages. It incurs a double violation of the human nature: privacy and social convention. One is to manipulate a person's dream, inside their own house and without their consent, fully aware of the perverse consequences of such an act. The other is to send to the victim's mind images and sounds of real characters (people in costumes) acting very badly — third-rate actors — in a studio that manipulates the ego in the dream. They stage soap operas that change according to the operators' taste. With this technology, it's possible to test all dream theories already proposed, from the strangest to the most cohesive ones.

Most of these dream soap operas are based on performances filmed live, always adapting to the physical and cognitive reactions of the "self" within the dream. Intellectual reactions are normally captured by SYNTELE, in which silent thoughts are also "uploaded" in the basic ego package within the dream and are always active and responsive — similar to their functioning in the waking state. A specialized AI can interact at a deep level with these more complex areas of cognition responsible for the interpretation of language and visual memory within dreams, helping the actors to create the next scene based on the interest behind the event.

As the frontal lobe is deactivated, the projection of motor activity — or the stimulus that would make the limb move — is analyzed by some D2K and Telemetric EEG modules (chapter 3) that capture the target's raw electrical waves. They transform this intention of the movement into numbers and parameters that creates a virtual projection of the movement that the target is trying to perform within the dream, compiled by algorithms that receive and process this data that continuously motivate them to respond "physically" to the stimuli coming from the artificial remote dream by the ego. This response can be assimilated by advanced algorithms derived from a "framework" under the supervision of submodules capable of using this sensory data and replicating them in a character inside the computer in the remote studio, thus creating a kind of neural remote virtual control. So, the event (the film being broadcast, shown and projected in the target's mind) is altered and this modification is constantly sent to the target's brain; such changes of scenes are adapted within the dream itself.

For example, the target is able to drive a virtual car in the studio computer just by reacting to the scenes projected in the mind. Each sequence of movement reflected in this car is updated in a period of milliseconds and returned to dreams where the projection of the car created on the computer updates its movement and adapts itself according to the target's motor reaction in response to the naturally acquired skill of driving. The report below demonstrates how this dynamic occurs.

The ego in the dream is sitting in the driver's seat of a car with the same perspective as that of concrete reality, which leads the target to automatically start the mental process of driving a car. To increase immersion, obstacles appear on the horizon and they may collide with the vehicle. These virtual obstacles have algorithms inherent in the programming of any game capable of interpreting and detecting the collision of the virtual car with the obstacles. Thus, the gamification of dreams occurs with modules in which advanced predictions based on machine learning can, over time, predict the attitude that the ego within the dream can take in a given situation. For instance, the instinctive mechanical reaction of turning to the right or to the left in the face of a sudden obstacle, or choosing whether to take a certain path in a fork in the road while driving the virtual car. This AI feeds on the data by capturing the kinematics of the movements ready to react to the intentions caused by the scenes, however, without the movement itself being executed. This prevents motor activity from "leaving the world of dreams" and expressing itself as a real movement, such as Belle and Aurora [30] did.

The individual practically plays a game in which they control an object thousands of miles away on a computer just using their thoughts. This object moves according to the target's intention based on their decision-making according to the possibilities presented by MKTECH. The projection, however, doesn't take place on a TV, but in the mind of the dreamer in a generally self-centered perspective. This basically means that it's possible to predict with great accuracy which trajectory the movements of all limbs would follow if they could be activated during REM sleep using only the electrical brain activity that receives afferent sensory feedback from visual feedback for dream movements as substrate.

The brainstem system periodically triggers the sleep state, during REM periods inserted in this particular brain configuration, in which the sensory input and motor output are blocked and the anterior brain — the cerebral cortex, which is the most perfected structure of the brain human — is activated and is bombarded with random impulses that generate sensory information within the system. At this point, the anterior brain synthesizes the dream from the information generated internally. It tries to do its best to make sense of the nonsense that are presented to it, even the images coming from D2K that take advantage of this innate configuration, literally hacking the physiological channel and generating severe psychological effects, including taking advantage — as we saw above — of these characteristics to gamify the dream with the ego inserted as the protagonist of an adventure in an old arcade game.

In the end, the ego ends up playing the role of an active agent capable of modifying certain objects, both in the dream reality and in the concrete reality, only through its acts within the dream. Although they are bits on a computer screen, this revelation is extraordinary and unprecedented in the history of modern science. The ego begins

---

[30] - Reference to the two monkeys that enabled the experiment to manipulate machines and mechanical arms with thought alone. It refers to the experiments carried out by the neuroscientist Miguel Nicolelis, due to the immense similarity between the two technologies. In this case, avatars move remotely only with the intention of the ego within the dream, as an extension of the limbs that are requested. The answer virtually comes with character movements on a screen.

to inhabit two worlds at the same time and carries the power to modify both realities simultaneously.

D2K has complex modules that go beyond penetrating our dreams and replacing them, always leading to a self-perception never conceived naturally. It has specialized tools to accentuate these characteristics in the form of games, as it makes them interactive. There are products available in the market to work with the raw EEG (electroencephalogram) data, used for dozens of purposes such as: controlling drones, playing video games, games that involve extreme concentration, working with experiments in laboratories, and so on.

The equipment that works with electrical signals is very advanced, but there is a gap between this old technology created as a military weapon and its modern versions used by non-military personnel (civilians). If you're curious, search the internet for terms like "mindwave mobile headset" and several technologies from different companies will appear as a result. This will give the notion of BCI technology with only EEG data that is available to the general public.

Operators have a location with plenty of space to produce dreams and send them via D2K. They create a true dream world in which the target is completely unaware of the reality of the waking and perfectly conscious world, while being fully awake to this artificial dream world where the focus is on the creativity for destruction. This technology opens a field of study for several areas of science. This module will be of great value when it ceases to be just a weapon and becomes a tool aimed at genuine scientific studies within ethical rules — and the law. This subsystem will revolutionize an entire study area, from personal traumas to remotely controlling a specific vehicle and the complete understanding of this process involving the mechanics of dreams in animals and humans.

The truth is that the brain isn't prepared to deal with these false images created in studio. However, the human being is extremely adaptable, so it's possible to reduce the damaging effects of this process over time, both asleep and awake. Nevertheless, even after a few years, dreams in the context of nightmares still mess with people's memory and perception of reality. Especially during the peak of psychotronic torture in which different situations are tested to infer the effect on the human subject and simultaneously determine the results of the interpretation of the generated situation, as well as the level of clarity of the events' details that was stored in the target's memory in small nuances, such as the simulation of colored scenes, black and white scenes, falls from a height, vehicle collisions, among others. Even the color of a particular scene is monitored to check the impact on the memory and the target's reaction to remembering the color scene later, equipped with a special RGB. At that moment, they collect logs — records of relevant events — based on the analysis of their behavioral feedback.

I remember a dream in which the target was sailing on a vessel. When looking at the port deck, they spotted the sea and it had extremely bright colors, standing out from the rest of the dream scene. The colors violet or indigo blue *hit* against the hull. The sea and some equipment inside the vessel were clearly captured by a 3D recording camera, mixed with scenes from a most successful RPG [31] game in 2015. These highlighted objects activate the interconnected module of the brain whose function contributes to a series of behaviors related to attention and the response to new stimuli. All of this happens in just one night, in an hour within the more than fifty thousand hours used in a systematic regimen of pain conducted by the operators in the target's mind.

Keep in mind that these electronic dreams are the daily dose of sadistic exultation by technology operators who spend 24 hours talking about how to make more striking dreams that deeply shake the victim in a new, negative way. This is one of the areas explored and with a high degree of interest by the operators behind the modern experiment called MKULTRA 2.0 — we're going to discuss more details in chapter 5. Tests of all kinds are created for their victims, real human subjects, without their consent and without being paid as they normally would if they volunteered for similar experiments.

---

[31] - An electronic RPG is a game genre in which the player controls the actions of a character immersed in a defined world that incorporates elements from traditional RPGs. It generally shares the same terminology, settings and game mechanics. Other similarities with tabletop RPGs are: the progression of the story and narrative elements, the development of the player's characters, the complexity and immersion.

The area related to dreams is like all the other modules in this book that require a great deal of knowledge on the topic. It'd be possible to write books and more books dedicated to the modifications of dreams given the large amount of information available. It could even include the study of normal dreams, which in itself is the subject of several theories and reviews. That is why I condensed the subject as much as possible in this chapter, keeping the parts relevant to the understanding of the general context of the technology and the production of dreams made in studio, as well as its consequences when processed by the human brain.

The quality of representations in dreams and the associations of direct ideas similar to those persisted in memory when passed on — when accessed in both realities — have the same degree of clarity and neurophysiological activation. They provoke the same feelings and emotions as an experience witnessed as awake — without the pain and some sensations linked to the executive faculties that are disabled — the subsequent impact and its other consequences are similar. The context in which it was acquired in the dream reduces the resemblance of mental memory to the physical object. Only by returning to the waking state and controlling the rational processes that we realize in which reality the new situations and the embedded memories arose; otherwise, it'd be impossible to know in what state this type of experience was acquired.

The synthetic REM dream is always displayed in maximum graphic and emotional quality. The clarity of its interpretation in our mental projection is infinitely more intense than the natural dream, causing a series of perennial changes in memories, in cognitive mechanics and in the set of intellectual systems. It can take some targets to a high degree of despair in which they're forced to walk towards the only viable solution to get rid of this invasion and manipulation: suicide.

Terrifying visual stimuli demodulated by our cortex during the REM dream activate the limbic system, causing the amygdala to be triggered, charging the system with fear and physiological reactions related to that fear, filling the body with stress hormone. Depending on some characteristics of the victim, such as age, the amygdala may stop functioning, deeply affecting human growth.

This could be the perfect way to study the dream and its mysteries, but it's actually used as a method of remote torture with serious consequences for the victim's physiological and mental health. What happens in the inner world of dreams can have physical effects on the brain and body of the dreamer. Such effects are no less real than those caused by the corresponding actions in the outside world — just the sensation of pain that isn't computed with the intensity that it would be, if it were real. The impact of certain dream behaviors on the brain and body can be equivalent to the impact caused by the corresponding real behaviors. Since these dreams are artificially created and read by the mind as if they were dreams naturally produced in the brain, they can have serious consequences for the victim's health, many of which are still unknown. Among the mental changes already proven, one that stands out regards attitude. Depending on the level of attack and the amount of energy used, D2K can change an individual's attitudes towards certain events, themes or people from real life.

Operators can take a particular subject in real life, which is inserted in the reality of the target, transfer that subject to the reality of dreams and start attacking it. This negatively alters the cognitive components that form the thoughts and beliefs about this particular object — which can be people, places, desires, vocations, etc. They drastically and subtly alter the affective component linked to that object by reversing its already conceived susceptibility of attraction to the feeling of repulsion, as it presents several impressive scenes. Thus, they modify their behavior both in dreams and in reality. This is yet another dangerous feature of this weapon.

Gap within dreams

"*Gap*" within dreams, or interval between dreams, consists of a tactic that can be mixed with other techniques, causing spatial and temporal confusion, as it induces the brain to try to contextualize disconnected stories that have major gaps between their *performances*. The ego takes the time to understand what is happening, as if it had passed out and had difficulty to assemble the puzzle of what happened after regaining consciousness.

A dream is initiated in a setting surrounded by forests with small houses in the middle of the woods. There are also native creatures that walk amid colorful and bright insects. As soon as the target becomes accustomed to that dream and interacts with the "creatures", something abrupt occurs and their vision goes completely dark. Some time passes by — the time within the dream is undefined; out of the dream only a few seconds have passed — and that blur disappears. The ego is then introduced into a completely different scenario, with real people, real houses and interactions between them. The ego is then inserted sitting in a circle in the middle of a conversation at a party where everyone wears a mask. Different colors and an atmosphere contrasting with the previous one confuses the target's brain. As soon as they settle in, there is another "gap" that puts them driving a vehicle at high speed through a muddy, rainy road. Instantly and unconsciously the target's mechanical reaction to driving is triggered. Then, the scenario changes again and the mind jumps to another dream. The ego is systematically *thrown* from dream to dream, one completely different from the other, until he wakes up feeling extremely tired. Some report severe hunger, due to the great effort and expenditure of the brain's energy in these events. We have several dreams inserted in one with no apparent connection in this "gap".

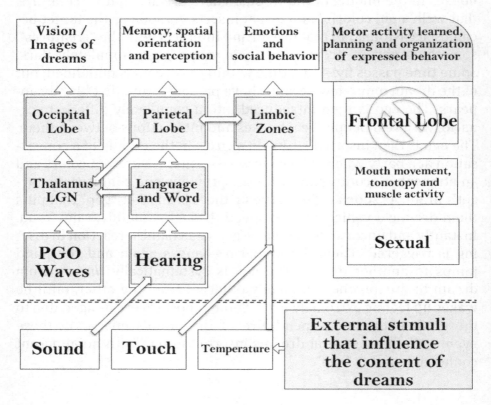

Figure 2.16 *A normal night's dream of someone who didn't have their mind kidnapped by the MKTECH scheme. Successful sleep is reflected in the quality of wakefulness. It renews us as does the act of drinking and eating that replaces essential substances. Asleep, one becomes unconscious and helpless. During sleep, the brain does a "biochemical" cleaning and eliminates toxins from synapses. It's an essential process that keeps us alive and ready for the next day.*

2.5.4 - How do memory, remembrance and imagination work?

We're going to take a break from dreams created in studio and replaced via D2K to better understand the influence of memory on dreams and on our mind. In this stage of the book, we're going to see the initial part on the subject — we're going to discuss in more detail in chapter 5.11.1.

Memories

Memory refers to the process by which we acquire, shape, maintain and access information. They're encoded by neurons, stored in neural networks and evoked by them and others. They're modulated by emotions, level of consciousness and moods. Thus, your emotional state, mood, stress level, attention and focus cause the memory to be registered and, later, accessed. These three parameters modulate the clarity in which memory will be "written down" and how the search engines will access it. We're literally what we remember.

The memory is not fully understood. For example, how and why some of them are erased and others aren't, in addition to the natural consolidation of synapse strengthening. Memories may result in a subtle synaptic change. These changes, in turn, can be widely spread throughout the brain, as they aren't stored in a central location. They are, in fact, distributed among several areas — a wise evolutionary strategy. If there is any damage to a certain region of the brain, it won't completely compromise memory and other healthy parts can adapt and be used as a memory area.

There are some distinct memories that have specific features:

* **Short-term memory** — memories that last for short periods. Memory responsible for remembering what you ate the day before, for example, and which will probably disappear from your database weeks later.
* **Working memory** — it remains available in the mind for an immediate response and is a temporary form of storage that requires systematic repetition to be consolidated in the mind, just like a phone number given to you orally. It remains vivid only if you think about it several times, or until it's written down and you no longer need to remember. These memories

remain until your goal has been accomplished. If it's useful, it can become long-term memory.

* **Declarative memory and procedural memory** — they represent systems that store long-term memories. They can go unused for years and be evoked sporadically to do, for example, crosswords. They're available for conscious access easy to form and easy to forget. Declarative memories. The non-declarative memory, on the other hand, isn't available for conscious access easy to form and easy to forget — implicit memory. Getting in a car and driving after you've learned how to do it is an example.

* **Long-term memory** — memories that can be remembered days or years after being stored, such as some remarkable moment in your life, some traumatic or intense event: a vacation trip or happy moments with your family or spouse.

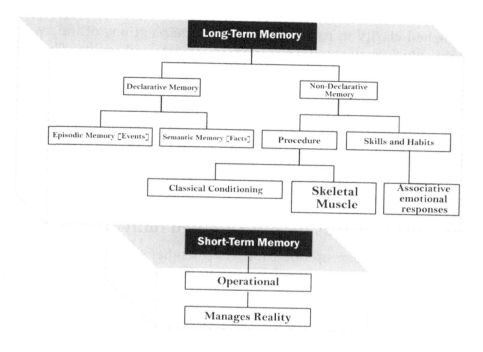

Figure 2.1.7 *Memories.*

Remembrances

When a person imagines an object, the brain presents a visual scheme that is very similar to what the real object is when the person actually sees it. The difference between imagination (or memory) and perception may only be a matter of degree, determined by the clarity or intensity of a sensation. In general, images and memories are much less vivid reflections than the original perception. Otherwise, it'd be difficult to distinguish the internal sensations from the external ones, as sometimes happens with those who have a tendency to hallucinate. Our normal ability to separate memories of past perceptions from perceptual sensations of the moment has a survival value. Obviously, we have inherited the ability to readily distinguish between internal and external facts, except within our dreams. There, emotions, remembrances and images get mixed up.

But this is because in the REM state the part of the brain that normally inhibits the vividness of the images is "disabled". This allows the memories and mental images to be released with a slightly diminished clarity in relation to the same perceptions of the awake state, which easily confuses dreams with external reality. The perceived degrees have their neurophysiological basis in the corresponding intensity variations of the neuron discharge patterns in the brain. Perceptual clarity is probably the main criterion used to judge how "real" something is. The brain is able to memorize information while we sleep. That is, we receive stimuli during sleep that can influence complex abilities upon waking up, and that will thus become part of the repository of memories available to be accessed during wakefulness.

II - Important tips for the Targeted Individual

Once you wake up don't try to remember those artificial electronic dreams in any way. You need to keep in mind that these are dreams totally created in studio, in an isolated location, by unscrupulous people. These dreams are different from natural ones, and aren't romantic, enigmatic, surreal and also have no secret meaning. Trying to remember dreams is relevant and pleasurable only when you aren't a target. When you're the "lab rat", in no way relive the dream in your memory. Operators know the perfect time to wake you up, as they monitor your entire neural network via remote EEG, and know exactly what will be noticeable in your memory. In other words, they'll always wake you up when something remarkable happens in the dream.

Since electronic dreams are more vivid than natural ones, whenever you try to remember them, the memories will start to get stronger over time. You'll reinforce the synapse connections associated with this false memory when trying to remember it. Then, dreams will enhance the suppression of recent memories of everyday events when you're awake. That is, you'll end up not knowing if the fact happened while you were sleeping or when you were awake, which would trigger the initial brainwashing process that is very exploited by the attacking operators.

There are certain artificial dreams in some victims that are so intrinsically strong that the target is no longer able to give due attention to the reality surrounding them while awake. They just find the time and focus to reexamine such false memories, unaware that they're causing damage to their brain and contributing to the results sought by the operators during the experiments. After a few years, it can make the target wonder if a particular event has actually happened. So, this striking synthetic dream may eventually become a false long-term memory.

2.5.5 - Creating long-term memories using D2K (BYPASS) by deceiving the brain.

Everyone dreams every night, but many people fail to remember such dreams the next morning. If you dream about superfluous or commonplace things, your memory automatically discards that stimulus. However, if the dream is very out of the ordinary — that intense nightmare that causes strong sensations, for instance —, it'll be registered by the brain. This premise of the mechanics of natural dreams is constantly used by operators. Now, at each sleep stage, remarkable and anomalous dreams that are reflected in all neurophysiological aspects of the target will take place.

We talked about several dreams in the previous topic, now we're going to approach a technique, a kind of highly damaging dream, which presents a clear connotation of experiment that occurs during REM sleep. They're used to create long-term memory straight to the unconscious state without going through the standard fixation process, and cause a huge mess in the process of fixing and accessing memory, which creates the "Déjà vu" [32] and "Déjà Rêvé" [33] phenomena, including the process that mixes both phenomena. Dream memories permeate the world of dreams and are accessed by reality, or vice versa, as we're going to see later.

---

[32] - *Déjà vu* can be described as a sensation triggered by a present fact that makes whoever suffers it strangely feel that they had already witnessed that specific situation, when, in fact, it didn't happen.
[33] - It's the feeling of *déjà vu* within dreams. Strange sensation of having already gone through a certain dream and evoked peculiar memories and sensations.

Mixing reality with dreams and dreams with reality.
Déjà vu and Déjà Rêvé

Faced with the possibility of creating any content for dreams, many procedures are being implemented in different targets, gradually permeating their lives as they serve as guinea pigs for modern experiments. In these experiments, they're completely unresponsive and absorbed in view of the richness of details and the level of clarity of the images and sounds that this technology is capable of generating in their minds.

The first times the individual comes into contact with these dreams, and their actions have physical effects similar to reality, they cause muscular tension, mental confusion, sweating and striking memories — perhaps the most striking they've witnessed in a long time. This continues to happen until they're aware of what is really happening or until they've gone through so many new experiences that the originality factor disappears. The target becomes "stronger", however, under the scrutiny of several scenes that will still occur until the electromagnetic waves cease.

One of the techniques exhaustively carried out on human subjects is created to specifically mix real images inserted in the dream reality combined with memories acquired while awake, overturning once and for all the fine line that separates the two worlds and transforming them into an interrelated reality. There is a procedure widely used by operators to insert false long-term memories generated in dreams without having to go through the natural memory strengthening process while awake. Its functioning is relatively simple: consists of accessing a real visual memory of everyday life stored in the target's brain — that may have been taken from a photo on the internet and has a connection with the reality, or may have been captured directly from the target's vision — and inserting it into the dream.

Let's take this as an example: the hallway outside where the neighbors' doors and an elevator are located. This common image is taken, a montage is prepared on the computer, and different people, animals and creatures are inserted in it. For a more realistic tone, a random person is also added, thus creating a fictitious situation with this real and everyday memory of the target. A dream can be generated in which a new neighbor — who neither exists nor lives in the apartment across the hall — actually appears leaving and entering the apartment. So, as soon as the target is guided within the dreams to access the corridor, that neighbor *placed* in the image of the hall will greet the target and maintain a short dialogue. Then, this scene is relayed during REM sleep several times, at each sleep stage. Once the target wakes up, leaves the hall and looks at the neighbor's door where the montage was made, this visual stimulus will automatically access the false memory previously inserted in REM dream, reminding them of the whole situation that was created in the studio and broadcast via D2K. This will produce profound questions in the victim, such as: "Is there a new neighbor?", "Did I greet him yesterday?" or "Was it a dream?".

This type of event occurs with many scenes of everyday life that are mingled with dreams. As the months and years go by, several memories like these will be widely used. Consequently, they'll be available when awake without the filter that determines in which reality that scene occurred, that is, in which reality that memory was acquired. They cause enormous confusion, permanent psychological problems, and emulate mental and behavioral disorders linked to memory.

This "cross-reality" technique is also largely used for brainwashing and for the creation of involuntary killers. The operators take a real scenario from the individual's daily life and make a montage with films. For example, the courtyard side of the building, the soccer field where they usually play, their school or college, among other diverse possible scenarios. Thus, there are memories of dreams recruited in the waking state and memories in the waking state recruited in dreams, generating "Déjà vu" and "Déjà rêvé" feelings in both realities.

The sleep state is considered extremely amnesic and paradoxical. It's a stage in which it is possible to make direct changes in human cognition, as it alters the unconscious data that make up the substrates of our consciousness and causes profound changes in the conscious "self". Dream processes are so important that they're designed to prevent us from remembering everything we dream of.

And this is usually what happens: we remember almost nothing from most dreams in REM sleep. It's different when you're connected to the MKTECH system. The Synthetic Electronic Dream — among other features — can make the dreamer remember practically all visual content and emotion, purposely made for the target to automatically replay the scenes in their memories upon waking up, thus embedding long-term memories as a consequence. A mental display of the content of the previous night is enough for this information, film, dream or game, to be remembered for many years, even for the rest of the victim's life. This further aggravates the health of the target, and intensifies the technique of mixing memories acquired while awake or in the world of dreams.

Another diabolical technique works with memories that involve some type of past compulsion that the target may be trying to get rid of: smoking, alcohol, illegal and licit drugs, gambling, sex, food. Dreams focused on a compulsion can activate the entire complex process that triggered these uncontrollable desires in the past. The desires, in turn, will surface immediately upon awakening, which affects the chemical functioning of the brain with a strong reminder of the reward and the feeling of euphoria and well-being that addiction caused during dopamine bursts. Let's take someone who got rid of the destructive habit of smoking, a very powerful addiction, one of the most difficult to quit that is linked to a daily routine and social and emotional components. During REM dreams, masked people — actors, real flesh-and-blood bandits — smoke or simulate smoking in an environment filled with a lot of smoke. All the details involved in a smoker's reality are emphasized, even the cigarette filter color will be highlighted, like a black and white movie with details that pop out and bombard the target with stimuli that refer to the old habit. Stimuli capable of making the victim relapse are inserted into the complex MKTECH system. This crime silently takes place and exploits all possible vulnerabilities of the individual, aggravating their abnormal behavior and causing withdrawal symptoms as the final result obtained in this distressing exposure.

It's a foul play by criminals behind this lethal weapon that slowly kills the target in an extremely painful and agonizing way. They also include addictions, compulsions and problems acquired during the target's life prior to the attacks in their strategic arsenal, turning such addictions, compulsions and problems into allies so that they can together achieve the ultimate goal of deteriorating mental states.

Creating the "Manchurian Candidate" [34]

This could be a research to improve cognitive abilities when correctly applied. However, there is only a behavioral change here — a change in the perception of reality — which mainly contributes to the Targeted Individual becoming an involuntary killer. D2K is vital to achieve this state of disgust and mental lack of control, caused by this devastating weapon. Manipulating the content that will be displayed in the target's mind during the nights is extremely devastating, as severe changes are created in the perception of concrete reality. All of this occurs through a professional narrative that synchronizes the theme of the day's attacks via Synthetic Electronic Telepathy with the manipulation of the dreaming mind, altering images and modifying memories directly. The two realities are integrated in a uniform way, reinforcing themselves without apparent borders and improving the absorption of the theme in the target's mind. This can lead the victim to commit acts of atrocity against other people, depending on who will be attacked and the level of stress the target is experiencing, as it uses the most diverse techniques that together can create remote killers ready to act without even realizing it. The procedures already described and those that will still come are part of this group that destroy a person's brain in a short time and pave the way for the creation of involuntary remote killers.

"Bypass" or alternate path to insert commercial images

Another very recurring dream with a commercial and thought experiment bias is to understand at which point in the dream a specific image transmitted via D2K will be absorbed and embedded in the memory in a more vivid way and with the highest degree of quality. The extremely disturbing phenomenon arises when a specific image

---

[34] - "The Manchurian Candidate" is a book written by Richard Condon about the son of a powerful family with a political tradition in the United States that is brainwashed to become an involuntary killer for the Communist Party. The work was adapted to movie in 1962 and 2004.

is sent seconds before the target is awakened. The desired effect occurs only when the target is highly immersed in the dream, which coincides with the deepest sleep stage. At this point, sleep is reverted to the waking state.

The conjunction of several parameters must be in sync for this phenomenon to occur, thus enabling the sending of the image at the exact point to cause the expected effect. It's so overwhelming that it's even possible to see, for a few seconds, the image traveling through optical radiation upon awakening. Meanwhile, the consciousness is gradually regained, and the target can momentarily see information traveling between realities, moving from the dream reality to the real life.

To determine the exact moment to attack the target so that the expected result is achieved, many intervals are tested and a maximum value is measured at a moment during sleep (and dreams) that indicates the greater possibility of a positive outcome of this experiment. Telemetric EEG data (chapter 3) together with the analysis of several other parameters related to dreams are examined to understand the whole dynamics of this process.

The phenomenon occurs until brain mechanics is reestablished in the waking configuration. The image at this point in the cortical space-time has the capacity to transform itself into a long-term memory, similar to the remarkable and traumatic facts while awake, such as accidents and scenes of graphic violence. The targets report that several experiments are carried out in this "magic" moment — in this unique window of mental manipulation. Operators upload images of well-known food product brands, apps logos or any other famous brands. These images remain at the end of the dream, and are perfectly vivid while awake for a few seconds. The most amazing thing is that it can be seen after waking up, both with eyes open and closed. The target can visualize the image in their mental projection with clear sharpness (even with their eyes closed), but they can't change it or replace it with another voluntary conscious visual thought. This final image is sent at the right time and the target is awakened from REM sleep at the moment of greatest immersion. Thus, this commercial image remains in the target's memory and will be easily available at any time for months to come. Commercial logos can be embedded in the mind in the form of visual memory. They can be accessed by the subconscious when, for example, a desire — such as the need to eat — automatically recruits the brand of foodstuffs engraved in the mind. This evokes the desire to eat through this demodulated image in a violent way within synthetic dreams. Such images have an unparalleled perceptual quality. They produce an impressive cognitive experience, a dream trip capable of leaving the target thinking about what happened during the whole day. The images reinforce the synapses related to the trauma, highlighting the memory acquired in dreams. These facts affect yet another level of remote experiments.

Over time, the *Déjà vu* ("already seen" in French) in the waking state becomes recurrent. Some places that the target never forgot — surrounded by strong emotions during the dream — can surface and create an emotional trigger in a waking state by provoking erratic behaviors and automatic mood swings. In apparently normal situations, on a walk, for example, the target may come across a real place that was used as a backdrop in certain films shown in dreams. Upon seeing the same scene, or structure materialized in both realities, unconscious physiological changes in the brain can be triggered, which dramatically affects the target's behavior in a sudden, abrupt way, and contributes to stimulate the electronic schizophrenia that we studied at the beginning of the book. However disconnected and impartial they may seem at first, these added details become the object of the experiment, known as **neural biological reprogramming.** The target's memories, actions and unique characteristics are gradually modified, affected and replaced to such an extent that their personality is unrecognizable.

Another macabre use of D2K is based on well-established horror film techniques, which vary in quality and plot. They can have footage similar to the low budget films of the 80s or even more complex films whose theme revolves around psychological terror. Remember an unforgettable horror film that you watched. These same techniques are used to shock, intimidate, frighten, scare and destroy the target's intellect as these terrifying films are created and interpreted by the mind. For example, the memory of the target's bedroom is used exactly from the angle at which they sleep in bed. In this case the bed is positioned parallel to the door so that a panoramic view of the entrance is possible. This image is then collected from the target's waking state visual memory. As soon as the individual falls asleep, a montage composed of the same image of the door is inserted in the dream — the entrance to the room was the last thing they saw while awake. Then they see what appears to be a relative with whom they live suddenly crossing the corridor. This person moves from side to side in a strange and repeated way, face hidden, until they stop at the door and look at the target lying on the bed. Unable to move, the victim just stares at the person. The relative with a disfigured face and unusual body posture starts babbling something terrifying with their voice completely modified in the studio to add more sensory data to the scene. It takes the terrifying atmosphere to a whole new level. The scene continues until the target wakes up in sync with the final attack. That is, when that relative runs towards them in a horrifying way and attacks them violently with no chance of defense, since the scenes that previously responded to the reactions of the ego limbs no longer do so. They're purposefully turned off to cause more distress. The perception that the attacker is a relative occurs within the dream with the use of the real image captured from any source. The voice, however, is captured via EMRa – Electronic Mind Reading (auditory).

The possibilities for producing dreams are endless. It all depends on the sinister creativity of the operators. There is no concrete formula for what works and what doesn't. This protocol is being developed right now by MKTECH worldwide. All targets connected at this very minute to this weapon and that are dreaming artificially are in fact human guinea pigs.

III - Important tips for the Targeted Individual

Whenever you're awake in a REM dream, in addition to not trying to remember the "dream", stay lying with your eyes closed for 10 minutes. As always, you'll wake up with a lot of V2K noise and great psychological terror with various facts of your private life exposed and synthesized; facts that were compiled and confirmed based on the reaction of your actions to the unfolding of the story in your dreams. Just keep calm. Think only of good things — in case your brain responds to the attempt to modify the current flow of thought, if it "pop the clutch". I'm not saying to ignore the operators' voices who are hearing your reactions inside your head, because this is a very difficult thing to do without focusing on something else. Don't let yourself be affected by this type of torture. Never give in. Morale and spirit fade each morning, as this weapon is made to lead you to death by complete starvation or debilitating stress, to blow you up from the inside out, to make you give up on life, to drain your happiness and vanish with any positive feeling, as it always explores the negative side in fake narratives and makes your mind work that way. Even to make you a Manchurian Candidate or end your own life. There are actually only two options here: give up, surrender and die or don't give up, fight and win! Always choose the last one.

Another ploy often used consists in external stimuli triggered by the microwave voice. In this process, it's possible to make the brain reinforce selective memories. This feature is used by operators to make the content of the dream more immersive with the support of sound that is responsible for contextualizing certain scenes or generating voices that will interact with the protagonist — the ego, their conscience — within dreams. After all, a movie with audio is much more captivating than a silent film. So, the brain also loses the ability to properly reinforce the important memories of the previous day and strongly reinforces the content of the dream that is being transmitted, creating the shortcut for building long-term memories that don't go through the normal process.

This mixture reveals a strong relationship between mental mechanics and sleep, when it's possible to bypass the entire internal system of fixation and assimilation of content. They reshape the brain as they shape play dough, using its synaptic plasticity, which is the brain's extraordinary ability to store information. The greater the intensity, the greater the frequency of action potentials. Thus, when arriving at the central synapses, the coded information is processed and modified in the neurons according to other information that is arriving simultaneously by that same neuron, coming from regions linked to perception, attention, cognition, emotion and many other functions — Synaptic remodeling, manipulation of memories in dreams.

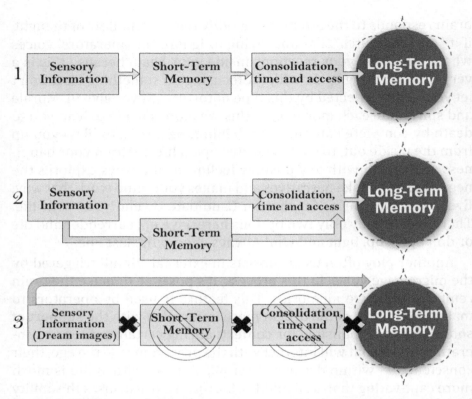

Figure 2.1.8 *Diagram showing how memory consolidation occurs while awake and how it's possible to circumvent this process with dreams created in studio and sent via D2K.*

1) Sensory information can be stored temporarily as short-term memory.
2) Permanent storage as long-term memory requires a consolidation process and can come from short-term memory or not.
3) "Bypassing" via D2K doesn't require any type of consolidation.

We know that people don't remember their dreams so clearly; a target, on the other hand, will dream and remember it vividly. D2K enables operators to completely control what is going to be displayed in an individual's mind. This becomes one of the most powerful weapons ever created by man within the Mind Control Technology system. It's undoubtedly the most powerful weapon in terms of direct manipulation of the mind, however, it works only when the target is unconscious, that is, asleep. It alters and modifies the entire cognitive process related to sleep and dreams, giving operators absolute control over the mind of others. Moreover, it's used as a weapon capable of waking the individual up and activating the sleep reversal process of modifying the thalamic rhythms at any time.

This technology opens up a vast field for the study of dreams in a controlled environment capable of collecting details never before seen in science, with possible and countless possibilities of discoveries for the humankind. Nevertheless, the technique of inserting data and long-term memories without going through natural processes — as seen above — is the most promising commercially. Technology operators are aware of that.

To achieve a long-term memory without going through the waking process, an extremely emotionally charged experience is required. Thus, the question arises: is it possible to place someone in a potentially dangerous situation in real life to create a traumatic experience associated with a visual memory, which leads to emotional levels capable of achieving this result, but without putting them in danger? The answer is no. It would be too reckless, something could get out of control, thereby undermining the person's physical integrity or even damaging the lives of those involved. In dreams, this doesn't happen. Any situation can be simulated with full control of the content. It's possible to insert a dangerous event similar to the concrete reality associated with several images that will be recorded in the long term without having to go through the natural process of consolidation of memory — through repetition of access and time. It's also possible to place visual memories quickly and "safely" in the dream, added to a strong emotional component that doesn't need to happen in reality. This opens up several possibilities to expand human capacity.

When the first altered dream in the studio invades the target's mind, it'll likely create an unforgettable memory. I believe that practically everyone who survives the attacks can report the point in time when their dreams started to become strange, striking and artificial. There are some standard dreams that are displayed to all targets, especially at the beginning of the attacks. A kind of primary program to be installed, a cinematic routine that contemplates dreams that will invariably achieve the outcomes already cataloged and expected on the victim. One of them is as follows: they create a massive creature that walks towards the ego of the dreamer and passes before their eyes in an undeniable splendor. This creates a feeling of omnipotence and omnipresence, and the target has a real feeling of being in front of a very powerful being. The image of this creature, the context and the associated emotions will become available as a long-term memory, equivalent to a real-life memory that has undergone a long process of consolidation. The sensation of being loved, the fear/insecurity, paranoia, disgust, horror, helplessness, incoherence and frustration can be felt with each passing dream, such as these.

Standard dreams are useful for accurately inferring, in the brain configuration, the nuances of behavior and thoughts associated and captured when reaching such states. They're then cataloged, thus knowing exactly on future occasions what sensations and emotions the target feels at a given moment. Both conscious and unconscious stages are analyzed. They assume that the unexpected and the extraordinary will always cause unpredictable results, as this is the most instinctive, organic and natural reaction possible that a human being can express in such an event.

IV - Important tips for the Targeted Individual

Operators will continue to attack your cognitive functions, even during sleep. You may be tempted to take the easy way out by taking strong medications to relieve stress and be able to rest again. As soon as you fall asleep under the effects of the drugs, operators will spend the night — as long as the effects are kicking in — using the dream interference weapon to create manipulated scenes that are even more vivid than normal. Since you're under the effect of drugs, the impulses that would normally wake you up — like a fall during your sleep or a scream in your ear — won't happen. Dreams can be even more haunting. The medicine will enhance the brain's ability to interpret dreams via D2K, and may have the opposite effect than expected: the famous paradoxical effect of the electronically induced drug. This can lead to total loss of sleep and may cause you to take more medication to return to your regular sleep, which unfortunately won't happen. The combination of these events will result in enormous mental confusion that can lead to depression or other problems, such an emotional experience never seen in a normal nightmare.

In one of these cases, the person in question didn't know whether they were asleep or awake. After taking the recommended dose, they went to sleep. They woke up in the middle of the night after some dreams that, among other things, reminded them of taking their medicines. So, they got up and took more drugs — all of this because they thought it was part of the dream. The next morning, the target who was unaware of psychotronic weapons, such as D2K, reported that they had the most horrible nightmare of their life. They didn't know they had taken five times more than the recommended dosage, as they thought that everything was part of a dream. The forces in the mind of these remote dreams are so intense the first few times that they're capable of causing deep effects for days. In this case, they almost induced the person to intoxication and death.

V – Important tips for the target

If you tell your doctor about your problem, he or she will prescribe sleeping medications at the very least. It's quite possible that they will also suggest hospitalization or something worse. I recommend you don't do this. It's preferable to try healthier ways to improve your sleep, such as physical exercise during the day.

Exploring the sleep transition state

The transition from the state of consciousness to sleep is filled with sensory mysteries in all aspects. This transition is intertwined with several psychological and physiological factors depending on the overall condition of the person. If they're under great pressure in life that results in constant stress and consequently a not-so-good night's sleep, it's likely that when they really need to rest after these episodes, the transition from wakefulness to deep sleep will come at an accelerated rate, predisposing the subject to events such as night bang (chapter 2.5.17) and episodes of more grotesque (hypnagogic) images.

With the detailed map containing each stage that updates the target's status every five hundred milliseconds, operators are able to monitor and interfere when switching from one stage to another. They create errors in neural synchrony that prevent the target from reaching the appropriate levels that represent a particular stage, reverting the brain's attempt to make the complete transition. At this point, dream illusions, memories and feelings of all kinds can be explored by operators.

Some consequences of these attacks combined with those that follow and precede the MKTECH torture cycle can cause a clear change in the synchrony between the sleep-wake cycle, and consequently create psychological disorders, such as affective and personality disorders. After a long period of exposure to this technology, due to the great emotional impact of this dream manipulation, alterations in the circadian oscillator can also happen. The permanent modification still needs further study. However, atomic hyperactivity, tachycardia, intense anxiety, rapid breathing, a throbbing head, headaches, difficulty awakening, slowness in reasoning, feeling unwell and sweating are the most common symptoms during and after the attacks.

There is also an attack based on transition states. It consists of sending back to the target only the emotion of the sub-stages of the transition state months or years later. Thus, the target relives that peculiar feeling without the need to reproduce the content of the dream associated in the past. Small electromagnetic stimuli containing the EEG configuration are sent in conjunction with sound and visual fragments, which stimulate the brain to access memory and continue the processes, thus reaching the objective of negatively reviving the previous state.

2.5.6 - Dream management using V2K

The human voice is composed of tones (musical sounds) and noises, which our auditory systems distinguish perfectly. The tones are characterized in terms of acoustic conditions by periodic vibrations. This division corresponds to vowels, such as tones and consonants (noises). A technique closely related to the brain's impressive ability to interpret the human voice during unconsciousness involves presenting, in REM sleep, an external stimulus as samples of consciousness, since the stimuli are incorporated into dreams. Temperature changes, internal organic changes, sounds and noises also reach the mind and end up reflecting in the dream in some way, whether in the form of images, sounds or both.

Almost everyone has experienced the sensation of hearing an irritating sound — a drill in the neighborhood, some hammering or music — and waking up a moment later to find out that what was really making the annoying noise was the alarm clock. Like music and other diverse sounds that can come to life within the dream when interpreted by auditory systems, the V2K microwave voice has a similar effect, however, it's more complex. The brain will automatically recruit the areas involved in speech and language to decode the meaning of that conversation and understand the message being sent.

This event — that incorporates noises captured externally by the dreamer into dreams — is purposely used by operators to conduct a narrative within the dreams, insert dialogues in characters, thus capturing the response of this interaction on the target via EMR. The innocent ego (without knowing they're sleeping) interact with characters that are part of their reality, that is, friends, relatives, spouses, family members, etc. The likelihood of this interaction generating reactions equivalent to the same scene, if it occurred outside the world of dreams, is very high. The ego acts normally in the face of situations purposefully posed by operators as they see the image — the face and voice of the character inserted in the dream in a way that is faithful to real life. Thereby, it's possible to discover the most hidden secrets of any person, enabling the creation of a response model analysis of the target considering a similar situation in real life, as we're going to see later. Even if the victim is able to control dreams — such as in a lucid dream — the day the ego fails to do so, will be the day that all confessions, secrets and hidden desires buried deep in the soul will be revealed.

Contrary to popular belief, we don't totally disconnect from the outside world when we're asleep. The sleeping brain maintains a degree of contact with the environment, as it searches for the meaning of the information that is received through our senses. We're able to wake up when someone calls us by name, but we continue to sleep if the name is someone else's.

This first experience showed that dreams could indeed be induced by direct verbal suggestion in REM sleep — to conduct immersive narratives, in stories with visual stimuli accompanied by a narration, which is similar to an announcer who describes a conjuncture of facts, thus placing the ego in the dream. Here, the aim is to discover details about the target's private life and everything around them. The operators are able to mobilize and recruit the attention and the inner forces of the human being, which causes severe psychophysiological damage depending on the content. One of them is to include deceased loved ones to interact with the "self" within dreams. They highlight negative points of the relationship between these two characters that occurred in reality — regrets, fights, bitterness, disagreements. The narrative and the "dubbing" of the characters within the dream with the image are united. As a result, extremely violent and negative feelings can be brought to the fore, altering the perception of reality and modifying the stages already overcome within the grieving process. It can even lead the target to commit suicide. If the voice that voices the character is used (which was previously captured via EMRa), the effect will be even more devastating and immersive.

Another important feature of conducting the narrative via V2K with D2K is to assist the invasion of privacy by supporting the REMOTE POLYGRAPH. Operators use the remote polygraph to forcibly check the veracity of all the target's thoughts. A kind of algorithm based on several parameters that automatically filter the thoughts and has three different possibilities: true, false or inconclusive. We're going to see in detail in chapter 3.

As the torture using the Synthetic Electronic Telepathy is based on attacking the target with incessant voices — which is naturally maddening —, the idea is to activate the target's negative native memories and dwell on that topic day and night. Well, if an important thought — of high value that intentionally violates the victim — has an indefinite status and raises doubts for operators, the doubtful thought will be replicated within their dream. Depending on the situation, it'll be re-enacted by the actors (operators), who check the target's reaction and point out whether it was true or not, thus resolving any doubt.

I know that it's difficult to reason at this unusual level of detail in an environment that has never been explored and that is really not part of people's daily lives. However, this is how this technology operates. The persuasive tactics and protocols of psycho-electronic wars implemented act in a sphere between subtle details, always crossing the boundaries of the concrete reality and the fantasy reality of dreams. We have to start dealing with this surreal novelty and we better get used to it. Reading this book is a good starting point.

2.5.7 - Sex-related dreams and testing of indiscreet situations

Sexual behavior is a vast, complex and exciting topic. It's one of the most pleasurable activities for animals in general and you feel better afterward. Sexual desire in humans can be triggered by erotic stimuli and external sensory stimulation, such as sight, smell, hearing, somatosensory system — touch, pressure, temperature — and direct stimulation of the genitals.

Neural control of sexual response in dreams works in a similar way to real-life equivalents. It comes from the cerebral cortex, where erotic thoughts occur, connected to the spinal cord that coordinates brain activity with sensory information coming from the genitals. This causes a reflex erection and an almost orgasm sensation, however, more subtle.

Sexuality is part of something great and is linked to the survival of the species. It's an extraordinary stimulus within a conception built on a strong psychological component established in structures, models and social values that govern and shape sexual behavior in the life of each one of us. Sex and its attributes that generate stimuli, impulses and preferences, create a set of characteristics aimed at sexual acts that also come to join the set of other active characteristics that make up the essence of what we call *ego*. This primitive impulse is loaded with character and personality within dreams. All of these characteristics related to the theme — visual stimuli and details related to sex, for example — can also be explored in the same way, just as other features within the dream were explored in the previous pages, thus forming a very reliable picture of the individual's essence.

To work with this topic, several aspects of the target's sexual fantasy are analyzed, such as the process of stimulation and "courtship" between the inner self and the mental images generated. Desires, attractions, even sexual thoughts are captured during acts in concrete reality when the target doesn't yet know about the technology that records their thoughts. So, the victim makes the most of their sexual privacy as any free individual on the planet would do, in situations of occasional daydreaming or reviving pleasant memories of previous encounters.

Operators calmly analyze each frame of thought, qualify small fragments and place them in a database so that, during the months and years in which torture will go on, they can insert the data accompanied by the sadistic episodes involving sex that will be created by them. These are the characteristics of the most elaborate use for psychotronic purposes, based on a feeling that carries the evolutionary factors that involve a great weight in all our complex social fabric and that are naturally explored on a regular basis in REM dreams and attacks via SYNTELE (Synthetic Electronic Telepathy).

Some sexual dreams are quite common in the life of a Targeted Individual, as they're used to test the victim's sexual preference and to know intimate details of memories related to relations from the past. The Synthetic Electronic Dream is a powerful weapon for this purpose, since it simulates any sexual encounter with anyone in any context. That is, it's possible to sexually stimulate the victim while they sleep, simulating encounters with partners, showing sex scenes between characters — some of these scenes can even be transmitted **live** to the target's dream —, sex between operators in their remote base (we're going to see details of these characters in chapter 5.6), in which the analysis of the term "operator" in more specific groups will take place, exposing their vices in human torture, their psychological disorders and a strong component linked to sex and latent voyeurism.

The target is then obliged to expose all their intimacies in an innocent or involuntary way within the scenes that unfold in their mind. These scenes are so real that they cause feelings of well-being and immediate euphoria when their deepest desires are satisfied. In addition, it provokes the feeling of invasion of the most intimate privacy of the human being.

Behind the events revealed in this category, a tendency in the experiments becomes clearer as the target daily embarks on these unconscious, synthetic and intrusive thoughts. One of them is to discover, within their sexual fantasies projected in dreams, a serious or hidden fact: betrayals, homosexual desires or paraphilia, such as pedophilia, zoophilia or any attitude that is out of the "ordinary". This serves to provide material for attacks on a daily basis, as well as to obtain a certain intimate secret and then blackmail the target with this information that can be sold to personal enemies or used to disturb the victim in psychotronic attacks that check their physiological reaction in response to this question captured in dream-based dating tests.

Several scenes of encounters that will end up in sex are presented to the ego in dreams in order to find out the victim's sexual orientation (homosexual, heterosexual, bisexual, among others). Actors generally perform the scene in front of a camera that simulates the target's point of view, mixing excerpts from adult films and other montages created in studio. As the scene unfolds, a virtual dream partner of the female sex is presented — if the Targeted Individual is sexually attracted to women. Then, the victim will have pleasant feelings and will let the "film" continue normally without objection and with a relative feeling of sexual satisfaction, similar to those of concrete reality, but less intense. If the preference is for same-sex partners, a natural dislike of the possibility of having sex with partners of the opposite sex will be detected and vice versa. In this case, the target's brain influenced by their sight — by the images in sequence that generate a film highly stimulated by realistic sexual scenes — can simulate something close to orgasm. However, the physical reaction (ejaculation) in men doesn't happen. The intensity of the act is attenuated. The sensation is more powerful and more striking in women though.

Most people have come across this type of dream in a natural way at some point in their lives. You know how pleasant it can be and how disappointing it is when you wake up, especially if it involves an unrequited love or a lost or never forgotten partner. These issues are widely explored by operators who play with the victim's feelings and emotions to discover their deepest and most intimate experiences related to the topic, later serving as ammunition for heavy bullying via SYNTELE.

Other tests consist of verifying the regions of the cortex that show responses given by certain stimuli that are still unknown to science — such as the orgasm and the regions that create this overwhelming feeling. Imagine the commercial potential for a technology that can, through electromagnetic stimuli, provoke the sensation of orgasm several times and without having to physically stimulate the genitals. Probably the founder of the company that launches the product will become a billionaire — which follows the clear line of commercial experiments illegally conducted on human subjects.

This technology is capable of increasing sexual stimulation or taking advantage of the so-called *wet dreams*, which are our minds' ability to naturally contextualize the involuntary sexual arousals that are triggered a few times a night. D2K, used to feed and escalate the immersion of feelings, is able to capture these nighttime arousals and indicate to operators that this phenomenon is happening at that very moment. The capture takes place by interacting with other MKTECH modules that, in the end, synthesize the information. It takes advantage of the natural erection that occurs in REM periods to work on sexual interactions, since the mind and body are in a module "configured" for sex. At that moment, the feeling of the encounters incorporates all these erotic data. Even "orgasm" can be felt more intensely.

Another type of attack aimed at the male audience is to maintain a strong erection by taking advantage of the target's predisposition to focus on sex — due to the lack of human contact for long periods, stimulated by an encounter with a likely partner or some event that is constant and more stimulating than normal. Live sex scenes, more intense erotic romance and libidinous thoughts involving the most diverse situations and fantasies in which the stimuli can be created in the mind aren't spared. Operators also take into account the deep prior knowledge of the target and information regarding sexual orientation that was acquired and refined during the daily attacks. As a result, the emotions created by intense artificial dreams may cause depression and an internal emptiness after waking up, for the target now realizes that it was all a manipulated dream. Undoubtedly, the protagonists played someone who might have a great affective/emotional bond with the individual, even if repressed.

"Dates" or pornographic encounters

In these porn-dream encounters that we're going to discuss in this part of the book, several devices are used to deceive the mental interpretation of external stimuli and signals, setting the ego in the type of event they're experiencing at the moment. One of the tactics is to use voices of people the individual knows — and that are stored in a database — to send sound stimuli during the dream, which provides even more immersion and realism to the scene. It contextualizes and informs the dreaming ego that the person in front of them is the same from reality with whom they had a relationship. A photo of that person's face is used as visual support, and makes the ego immediately identify all aspects of the scene and thus relax and interact perfectly with the fictional representation. The victim is carried away by the narrative and usually interacts in an approximate way of how they would do with that same person if this happened in the concrete reality. Keep in mind the amount of personal information about the target that can be obtained by sending their ego into stories that will reveal practically everything about their emotional and loving intimacy, desires...

Now imagine that the Targeted Individual has a longtime friend. They've known each other since they were children and cultivate their friendship to this day. If the operators want to know if the target has already had any intimate contact with that person, if they intend to have or actually have some deep feeling for their friend that wasn't detected when capturing the data during the day, technology operators will get images of that friend during the night. These images can be taken directly from the victim's visual memory or from a photo posted on social media, and they use pre-recorded voices of that same person from their database acquired via EMRa. Thus, unconsciously, the voice that is sent via V2K will be quickly recognized by the individual that will feel safe and immersed in the scenario created by the system. A montage that, at a given moment, emphasizes a photo of the friend's face is enough for the interaction with the rest of the film to unfold. Based on the target's feeling reflected in the ego, transforming into sensations felt by the mind and body, which will automatically interact with the virtual friend in the same way as they would in real life, or at least very close to that. The "grandmother cell" and the complex specialized chain linked to human facial recognition are activated, automatically responding to the face presented within the dream. After the ego is at ease and interacting with their "friend" within the dream reality, actors will make the show happen.

One of the applications in the use of this weapon — that stood out among the many observed — is of a clearly illegal human experiment disguised as torture. It revolves around the feasibility of carrying out dreamlike pornographic encounters between the ego and the virtual projection of the operators after being filmed, digitized and sent to the target's mind. This scenario opens doors in several fields, for example, studies for scientific and commercial purposes with the clear objective of promoting encounters between characters in dreams. They may be two dreamers (egos) in a community virtual reality in which the reaction of a person influences the other, which in the future will revolutionize technology.

This is one of the reasons that corroborate the assertion that every target under attack by psychotronic weapons is a test subject for projects like this. They employ this whole study in the development of a technology similar to a dating application, but using only the ego in the dream. And they're getting closer and closer to achieving the viability of this macabre "app". I'm going to explore this topic in depth in the last chapter of volume 2.

It's impossible to describe in a few pages the feeling of having these dreams/interactive films of sexual encounters. It's extremely realistic. The interactions that respond to the ego's commands within the dream make the dreamer really feel the intimate contact with their partner. The pleasant feelings and physiological reaction resulting from this contact that occur in the mind and parts of the body are almost the same if they happened in real life.

Inserting familiar faces in dreams

A feature that is overexploited by operators is based on using photos of familiar faces that the individual — while dreaming — quickly recognizes. Usually, faces that appear in dreams can be very diffuse. An automatic visual recognition isn't possible. However, when we come across familiar faces of friends, relatives, spouses, famous people and even movie actors in our dreams, our automatic recognition mechanism works this information unconsciously in autonomous mental processes and in even deeper layers of the human mind, close to the sensations that make up the substrate of disordered and raw cognitive material, to later present the result to the ego.

Just like in real life, the ego in the dream goes along with this built-in and activated facial recognition feature. Taking advantage of this, several cinematographic situations have been created and used extensively with hidden intentions at first sight. In dreamlike sexual encounters, when operators want to simulate an actual interaction, such as the previous illustrative example about the target's friend, a real photo of that friend is taken, either from their visual memory or from internet photos on social networks that are abundant in this era of virtual narcissism. This photo can be pasted on a body in any montage — it can even be attached to the actors' bodies in the dreams. This combination will be used with systems similar to those available on the internet that calculate the trajectory of the body to be replaced in the video, thus replacing the actor's face with that of the friend in question. It only takes a brief appearance of that face in the dream to begin the process of facial analysis, which instantly promotes comfort for the ego. These sex scenes occur due to burning desires, in some cases they generate all kinds of stimuli. It's indeed a great "trip" in the act of having sex almost completely. In these cases, the microwaves voices also reinforce the auditory recognition of the person in question: "Look, it's me, your friend. I miss you...".

In addition to testing various parameters of the technology itself and the behavioral consequences of the ego, tests on intentions in relation to the person the target was related to in dreams (who is a real person in concrete reality) are also created. The use of the face is implemented in several conditions, not just sexual encounters. It's systematically applied in almost all contexts, including to make the most of the torture and to cause greater damage during D2K attacks with more realism involved in the narratives and the overwhelming emotions reflected in the target.

The so-called visual memory is recruited when a face is presented, and is activated by a set of neurons in the Temporal Lobe (Inferior Temporal Cortex) responsible for the management of inferotemporal memory that is linked to facial recognition and several areas distributed by the responsible cortex by sight. Facial detectors are located specifically in the inferotemporal cortex of both cerebral hemispheres. Neurons in the mid-lateral and medium-deep areas can become active when a face is facing you; others can be activated when a face is on the side. In the anteromedial area, neurons react to the presence of the faces of specific individuals.

Keep in mind that the brain works with the concept of distributed memory. It's a clever way for nature to defend itself in the event of a catastrophic system failure, death of neurons or accidents, and yet memories may still be available. If the electronic dream presents people's faces, the cortex responsible for the visual comparison of facial recognition will be automatically activated. This can accentuate the recording of a false memory that will be inserted directly as long-term memory. Thus, the autonomous ability to recognize faces recruits a range of brain resources for this purpose. Subsequently, upon recognition, another sequence of internal phenomena will be triggered based on several aspects.

As soon as your friend's face is detected, luggage from the past in terms of memory will be recruited and feelings about the virtualized person will accompany the post-recognition, such as the target's last impression of the person, the synthesis of the analysis that a person perform about someone else in terms of appearance, lingering feelings, misunderstandings, jealousy, emotional peak, opinion about conversations, personal life, pleasant or unpleasant acts, relationship status, if something negative happened and left some bad feeling or a positive shared experience, if some emotional defense system was activated, discomfort or other factors. All of this stems from the current situation of the relationship between the target and the real person projected in the dream. This feature is compatible with any recognizable person in the target's social circle portrayed within the dream. It'll likely produce these aggregated results right away. So, it's possible to know the target's opinion about everyone and the reason for that opinion.

A shady capture that runs behind the scenes occurs just between the point of facial recognition and the creation of the image model that one has of that person based on conscious and unconscious factors. In order for the brain to recognize a face, it analyzes a number of unique features, such as skeleton shape, distance between the eyes, surface texture of the face, color of the eyes and skin, and specific contrasts in different regions of the face. Furthermore, this mechanism activates other areas linked to the brainstem, which is responsible, in addition to the alert and wakefulness functions, for recognizing the physiognomic alteration of the affective state: anger, joy, disgust, sadness, and so on. This data will be captured and, after the sexual encounter, will be debated via SYNTELE (Synthetic Electronic Telepathy) while D2K keeps the artificial dream running. They extract even more information about the target and the dynamics between the projections within dreams. A very clear picture of their sex-related social behaviors is then put together.

Some considerations are somewhat peculiar and the exploitation of the post-encounter is one of them. Let's say that the target's friend (the real one) who was used for a dreamlike sexual encounter is married or newly separated and their husband is a great friend of the target in real life. As soon as the sexual intercourse between the individual's ego in the dream and the projection of their friend ends, an operator's voice will ask in a provocative tone: "Now, what are you going to do? You had sex with your best friend's wife!". A test on this possible scenario will automatically take place, which produces data for future analyzes of the target's emotional reaction — even their character can be tested. Post-encounter feelings reveal even more details of the victim's personality and how they face certain situations involving sex and people from their life. These tests regarding the target's moral and character occur amid sadistic and cowardly enjoyment by the operators behind the technology, followed by intense physiological wear during the act and upon waking up. In that case, it's a bit of a relief, but it all depends on the target, of course.

Several scenarios like this occur daily for countless purposes. The most harmful of them is the inclusion of false memories, brainwashing, distance from reality and the increasingly deep harmful immersion with MKTECH and its "virtually" generated world that destroys cognitive processes and corrodes from the inside out. Kill the brain and the body will perish too, but first inflict the greatest pain in the process and create data for the experiments as well as fun for the operators and leave no trace. This is the motto.

REM rebound

Another tactic used on victims is to wake the target whenever they go into REM sleep. As soon as the electrical configuration indicates the entry into the stage, the victim is awakened with screams or noises that echo in the mind. It's an act that occurs systematically until the target is able to sleep soundly from exhaustion after days of sleeplessness. This strategy — combined with the stress experienced during the day caused by SYNTELE and the difficulty in sleeping — provokes an effect called REM rebound. This effect alters the regular regime of dreams, which become longer and deeper within the REM in proportion to the duration of their deprivation.

Thus, long-lasting dreams are generated with a lot of debate between the operators and the target involving very realistic interactions. An entire soap opera unfolds around the characters. The target, who is the protagonist, is faced with a complex world created artificially. This event usually takes place after a massive attack at night that prevents the target's transition to sleep, and the process is only triggered in the face of physical collapse. It's usually preceded by the well-known "Swarm Attack". This attack consists of a full force attack that uses all available infrastructure and personnel resources over a long period, as we're going to see in detail in chapter 5 of Volume 2.

While the body rests heavily from fatigue, dreams that are more like an episode in a series interact with the ego in several situations of intense dialogue, which profoundly alters memory. They cause disorders and cognitive changes that permeate all aspects of life.

2.5.8 - Creating test models about the target

In addition to creating sexual scenarios in dreams and "playing" with the target about their deepest intimacy, desires and their mental stimuli to maintain arousal, for example, operators can also evaluate several situations in which there are countless comprehensive models of various qualities or weaknesses, their character and psychological profile, how the target deals with decision making and intentions on a given hypothetical scenario. It's even possible to test the person's character, whether they are honest or not. Aspirations, repressed desires, regrets, among other peculiarities. To discover something about the person is easy when using an artificial dream previously created and filmed with clear situations and events containing a defined context, as already discussed in this chapter. The target's responses to certain circumstances say a lot about the target itself. Operators then create rehearsal situations focusing on future projections inside and outside the dream world.

An example is to test the victim's character in a situation that resembles a real scenario when they're dreaming in REM sleep. For instance, they produce a scene in which a huge amount of money in bags is placed in a specific context. In this scene, the ego is urged to pick up the bags by people who look like their cohorts in the dream that starts from that point so that the person isn't influenced by any external factor. The victim can only use the resources already available of their strength of character. If the act of stealing is morally inappropriate under any circumstances, the target will automatically reject the bag. However, if they're coerced to steal, they'll feel a clear discomfort afterwards.

This entire process in dreams is closely monitored by operators who receive continuous mental and visual feedbacks from the target along with vocalized thoughts for verbal communication regarding the actions the ego will take within the theatrical scene and, most importantly, the intentions, second intentions and why operators want a certain reaction. All of this can be placed in a database in which complex statistical analysis compare a behavior in the same situation within different moments when the target's behavior was completely different, including their growth and their learning in the face of the attacks. In other words, the way the victim behaves in different situations and similar scenarios that were created to induce certain behavior being tested. So, the AI is able to predict the likely behavior in some similar event happening in the common reality with great accuracy, and to group unique and distinct characteristics of the target's personality together. Few points outside the curve are detected in these processes. In this way, this model makes use of the "conversation" with classes or modules of other MKTECH technologies, consulting, computing and measuring values. For example, confirmation by remote polygraph masterfully indicates that every decision or thought on a subject is a likely lie, or truth. Remember that people are different. Therefore, there are people capable of lucid dreaming who can mislead the whole process in the chaos of artificial dreams. However, I didn't have the opportunity to search for someone with the ability to be an oneironaut. The question remains open for analysis.

Some points are necessary to form an accurate analysis of all aspects. Models that test a person's honesty, their sexual orientation, fears, virtues, character, whether the person is ambitious or not, their concerns, desires, nature, moral circumstances, real impressions about events or people, virtually all types of experiments can be conducted on the target in a compelling way. They deceive them and make them believe that they're interacting in concrete reality, but they're actually within a dream in which operators can change reality in any way that is convenient for them. No rules, no laws, no limits.

Attitudes towards feasible scenarios in dreams are extremely important. After all, they reside in the fact that behavior is, in general, caused by a set of skills and feelings. By knowing the attitudes of someone in dreams, one can predict their behavior in reality with some certainty.

Other common tests consist of mixing several techniques that we've seen in this chapter. It's worth noting that the operators conducting the torture and the experiments also adapt to the target's personality. Each day they must present a new scenario to be explored, thus draining possible resources, and always reporting the machines and the other layers of the experiment. At the same time, they undermine the target and drive them insane, in the hope that the status of remote killer occurs before their death.

Taking into account all normal states of consciousness that an individual goes through every day and night, the state that has the widest range of physiological variability is the REM sleep. It can take the target from heaven to hell in seconds. By taking advantage of this fertile and not yet fully understood ground, other procedures are carried out in the target's mind. One of them is based on techniques that mix the target's memories of events that occurred in the past. It recalls an atmosphere of a happy time in a house that has been recently sold. It also includes a couple of dogs who were loved members of the family for years, but that have recently died. The insertion of these positive images causes comfort in the victim who relives the golden times of their life in a pleasant way. The calming sensation of running the hand through the animals' fur, searching for fruits in the garden, the laughter and happiness from the past are relived with a magnitude of intense response. Suddenly, the target is taken to a degrading condition. The death of their beloved dogs and the loss of the house where they'd like to spend the rest of their days are inserted into the context and generate a charged, negative and violent emotional state. This dream is repeated several times with small variations. Other losses and defeats may be explored as well. Thus, the repetitive sensation of irrevocable trauma is used, contributing to the malfunction of the systems as a whole, which corrodes and intoxicates. All of this negativity is taken to the waking reality that destabilizes the entire psychological process, without taking into account the physical damage that these actions cause, such as erratic electrical bursts in which the REM dream remains latent even in the waking state for a few minutes.

The target's concerns about their daily, family, financial and existential life are also explored. Concerns about the attacks begin to be discussed and start to dominate in dreams. This theme is used to instill, through brute force, a parallel reality — neural reprogramming — that forces and directs the focus of the target's acts only on the content that is generated in their mind and that occupies a good part of their cognitive capacity to the detriment of those that are really important and that arrive every second in the target's daily reality. Various techniques for inserting memories and emotions are initiated.

One of these concerns was captured and mixed in dreams — it's usually highlighted and easily accessible. Once, the target was extremely worried with the course of their current job and, during that same period they were feeling homesick, as they had recently moved. This concern was captured and exploited by operators who created a fake invitation to an interview. The individual dreamed of the place, a beautiful striking landscape. They admired a glamorous sun that faithfully reminded them of their hometown atmosphere as they made their way to the interview area. At this point, the assimilation of the dream and the connection to one of their biggest concerns converged. They thought they were actually heading for a new job interview. This disturbing assimilation — and the world carefully created by operators who led the target to reach this conclusion during the course of the dream—, is impressive and provokes a feeling of pride and satisfaction for the automatic receptivity of the visual context linked to the mitigation of their main concerns.

A few words via microwave voice always help in this regard. "Your interview", "Great job opportunity in your city", the enemies uttered, and the target believed. "I went to the interview" and "I passed". Soon after, under the weird creativity of an operator, they used memories of a floor of a well-known mall, where the target started to live. Without privacy, escalators brought in passersby, who were eager to shop, from various regions. Suddenly, there is a fight. Chairs and tables are turned and everything is broken on the location. End of the dream! The target wakes up disturbed by a volley of insults (via SYNTELE) in their mind in conjunction with the sad realization that they were once again inside a manipulated reality that was happening directly in their mind with subsequent effects in the rest of the system and body.

In the first year of the psychotronic attack, D2K basically makes use of a script within a catalog based on a wide range of primordial pre-programmed nightmares. These nightmares will result in known effects in a uniform and linear manner in most of the affected targets, whose expected results shall be listed. The intention is to take advantage of the unprecedented nature of the event and the way it affects the brain, thus exposing the victim to primary conditions that reveal their internal reactions in total bewilderment. First, they run the basic attacks in which most first-time victims will suffer the consequences of those dreams that modify the memory and target the brain, such as reactions to sexual encounters.

Those stimuli — in which the reaction depends on a largely random, reflexive process and the reflected consciousness — can provide specific information that will be used as the primary skeleton of an absolute representation of the "self" in the future. From there, they create intention and rehearsal models, both in reality and in dreams. We can affirm that the two realities in these models, both palpable and dreamlike, start to interconnect — within their differences — and practically forms a continuity in space-time.

2.5.9 - "Tunguska Sound" or Night Bang

What is this event after all?

Tunguska Sound is a type of episode that only happens when a person is at the beginning of the transient state of wakefulness to sleep. It's kind of unique. Apparently, it just happens in a determined configuration of brain waves in a certain physiological state. It usually occurs when the victim is extremely exhausted, stressed and without adequate sleep for weeks due to the constant torture caused by psychotronic weapons.

The impression that one has about "Tunguska sound" is that someone set off firecrackers next to their bed, or that something that resembles a shot was fired next to their ear. After that, the target feels as if the sound is "bouncing off the skull" and echoing in their auditory systems. This reverses the sleep process immediately, causing the target to wake up suddenly and stupefied, discouraged and frightened by the violent bang that only they can hear. The bang can be generated by a sound that is overpowering, which was demodulated from a carrier wave of any nature, in the low or medium frequencies or a high-pitched shriek.

Some questions are promptly raised on the subject. For example, what conditions are capable of making a person vulnerable to such an attack? What neural, electrical and physiological configurations cause this type of phenomenon? At what stage of deep sleep is consciousness at? Future studies are needed to get such answers.

What is strange is that, in order to feel the pain in the eardrums, the stimulus would have to come from the displacement of the air in a determined intensity in the form of sound. However, the Tunguska Sound is characterized by an explosion from the inside out, caused within the dream by an internal stimulus and without the external sound impact. In the field of speculation, operators must exploit a flaw of a mechanism known as "neural adaptation of volume" for this phenomenon to happen. This mechanism consists of the natural cerebral reduction in the auditory sensitivity of external and internal noise that occurs in one of the sub-phases of transition from wakefulness to sleep. This happens because the perception of mental images and internal sound activates the same areas in the brain. This feature may not be available to be used at all times by operators. If they could, they would use it without moderation.

Once, a given dream on a target caused a huge headache and the sensation that the auditory systems were damaged after the shrill scream demodulated from the V2K attack and executed in the correct electrical configuration. The sound caused the physical sensation of damage to the auditory system when it was interpreted. The impression was that the ears would "jump out of the head", such is the power of the perception of the event by the conscious ego within the dream. The pain itself wasn't really felt upon waking. However, the beginning of the action was triggered, the conditioned reflex of the organs involved in the act. The initial sensation of this impact was captured — the area was throbbing. However, the somesthetic area hadn't been activated; only the memory of a ghastly cry inside the mind along the lines of the "Tunguska sound" remained. This time, the consequence of the "impact" that this cry created in the ear canals was felt without the actual sensation of pain upon waking up.

2.5.10 - Dislocated Thought or Dissonant Mind

This other phenomenon occurs depending on a set of physiological and neurological factors added together, usually during the "Swarm Attack" period (Volume 2), in which sleeping becomes a privilege. When this occurs, because of all organic circumstances, the target falls asleep due to fatigue. At this moment, with all MKTECH equipment working at maximum, the extent of the sensation of the images interpreted by the brain in such conditions is increased. Let us say that the immersion causes vivid experiences composed of interactions from different sources of projections of the interpretation of V2K, SYNTELE, D2K, and the elements generated converge into an almost illusion composed of distinct feelings.

As they proceed in deep REM sleep, the target will invariably wake up at the emotional peak of dreams and at the exact moment when the brain is already working to its fullest in deep sleep configuration. As we well know, operators have complete control over all aspects of the sleep/wake process. Thus, when the artificial dream is at the height of immersion, the regression to the waking state will abruptly happen. The dream will remain "running" in the target's mind for several minutes, as it persists in the REM state. This phenomenon occurs naturally in humans, however, without the same intensity. The first symptom is the physical resemblance to a hangover. The second phenomenon occurs by increasing the natural effect of changing the brain's neural configuration. It takes a much longer time to resume the configuration of brain rhythms. In this terrifying phenomenon, the attempt to think consciously — that is, rationally — can be experienced, and the brain may not respond to the request. At that moment the mind didn't "reconnect" with some more complex cognitive functionalities that are only accessible in the face of total consciousness. This includes more refined functions, such as executive ones. Highly charged memories available during dreams haven't yet reversed conditions. The world that took place in the reality of dreams still permeates neural processes.

The refusal to respond to the rational request to change the content of thought — that is projected in the mind — provokes a strange feeling of dislocated thought or dissonant mind, in which one imagines, in an abstract and philosophical way, when the mind is reconfiguring itself. It resembles an object out of place, detached, with no connection to the conscious "self". As soon as the request isn't executed, an error is returned as a response, a kind of bug. There is a flash in the background, neck and head tremble accompanied by a feeling of anguish, twisted images and a lacerating emotional pain are then dissolved in various effects resulting from the interpretation of these responses, or the lack of it, followed by an effort to regain access to functions still not completely reactivated that remain working with the dream reality. Thus, the attempt along with the natural recovery of the conscious is gradually making the brain and the mind *boost* after several attempts until its correct functioning starts normally. Some problems of memory indexing and ambiguous, nostalgic and depressive emotions will occur.

Within this same attack, it's also possible to create neural configurations by electronic stimuli which, when abruptly bringing the victim to the waking state, immediately awaken them up with feelings of loss and anguish and a unique feeling of emptiness already *installed* even before the target fully regains consciousness. This modification occurs even without the graphic use of dreams or the influence of manipulated images, which shows us once again how dangerous and powerful the weapon is regarding the modification of the human psyche while sleeping. It works in layers of the subconscious, modifies the natural automation of processes that act outside the conscious mind, and cause serious problems of artificial changes in the formation of thought and feeling. I venture to say that such obscure, intense sensations are changes in the chemical and electrical patterns of certain areas of the brain, based on a previous mapping captured when this feeling has already occurred in a state of consciousness. Thus, they force the same state — or part of it — moments before the target wakes up with some kind of signal that doesn't carry the images that make up the artificial dream. This signal is demodulated by other regions of the cortex causing these profound sensations.

It's not yet known what this signal is and how its content is capable of affecting complex social feelings in this way. I calculate, however, that it stimulates these states with something similar to composite signals containing some type of noise and other specific content undetectable by other areas of the brain, capable of being demodulated by specific regions of the prefrontal cortex that is responsible for, among other functions, orchestrating conscious thinking and acting in part of the limbic system — responsible for emotions. Nevertheless, we'll only know for sure when all the parameters of this technology come to light.

2.5.11 - Altered time perception

Infinite cinematographic techniques can generate effects — that are still unknown — with the use of this weapon, however, something caught my attention: another type of attack that occurs using a cinema feature. This feature consists of showing the movie (dream) in an accelerated mode (from 2x to 8x) in the mind, activating other weird sensations related to the effect.

This accelerated dream is produced to sustain and accentuate a unique sensation that becomes predominant in an intense night of attacks in which everything seems to be accelerated. Even the perception of time in the waking stage seems to be affected. It usually occurs due to acute stress and the constant maddening hysteria. Consequently, it keeps the target extremely alert even in a sleep state. The effect of presenting the dream at an accelerated speed ends up creating adverse reactions, such as nausea, dizziness, headaches, as well as the delay in regaining consciousness, which aggravates the victim's dissonant mind.

2.5.12 - But, after all, how is it possible to replace dreams?

Normally, the conscious mind acts as a computer "firewall" to filter out destructive, invasive and negative ideas that begin to form through external influences. We're able to organize these external stimuli, control our actions, manage the image that will be displayed in our mind and regulate countless aspects that together we call the "self", our consciousness. Therefore, while awake, it's not possible to insert mental images. They're in fact meant for the dream due to the difference of several characteristics and configurations in neural states.

Primary activation of the central visual system is one of these factors as it has "preference". Visual conscious awareness is taking up a large part of brain processing — image analysis, facial and spatial recognition, memory, among others. Even if the target closes their eyes, it'll still not be possible [35] to capture and demodulate the waves that come with the invasive content embedded via D2K, as the mind is still working in the waking state.

So, for now, brain and mind cannot be completely dominated by MKTECH invaders, as they're awake and their conscience, their "spirit", is in control of the situation. From the moment we start to fall asleep (as we become unconscious), there is a decrease in the firing frequency of brain stem modulatory neurons, and thalamic neurons indicate the change of stage. Neurons — whose axons branch profusely — transmit generally modulating information. They participate in the sleep and wake control system, transmitting generic information to the entire cortex that makes it possible to turn on, amplify, attenuate or turn off brain activity as a whole, just like a dimmer that changes the brightness of a room. Therefore, the indication that we enter the sleep state occurs and other settings inherent to that state are ready.

We know that dreams are basically formed by visual memories and images previously stored in the brain. As the system that sends signals from the retina to the cortex is momentarily deactivated, access to the images, their display and interpretation by the visual cortex is left to the areas responsible for accessing such memories. Here comes the **Thalamus**, more specifically **the LGN or Lateral Geniculate Nucleus**, which is the gateway to the visual cortex. The funny

---

[35] - There are more advanced weapons that are able to insert visual thoughts into people's minds similarly to D2K. Even if the target voluntarily tries to stop them, by closing their eyes and being conscious, it's possible to see the invading film starting in the face of the visual thoughts. In order not to rush, we're going to talk about this weapon on another occasion.

thing is that the retina is not the main source of synaptic input. 80% of stimuli, excitatory synapses, come from the visual cortex in a feedback process.

As these images travel in a feedback and bidirectional way, they supply the thought in sleep with its characteristic visual content, executing the normal flow to compose what we call a dream. During the feedback process, when traveling through the LGN to access visual memory, loading the signal with information and returning it to the cortex for interpretation, D2K sends its resulting signal ready to be interpreted by the brain at a frequency that replaces the natural signal of communication between the LGN and the visual cortex. In this way, the false dream is received by the cortex and is normally interpreted as if the content was legitimately loaded with the images stored in the target's brain. As more content for the dream is requested, the hacked LGN continues to send this fake content without the visual cortex realizing that it's non-native information. This image sequence triggers the whole process inherent to natural dreams, such as the use of limbic systems of the amygdala, the hypothalamic functions that release hormones or induce — in the image, scene or character in the target's mind — an emotional signature that deeply modifies the process of storing information in the mind.

The LGN not only controls the flow of information from the retina to the cortex, it's the first place where what we see is influenced by the way we feel, shaped by the internal state, and attached to the image recorded in memory. It's also argued that the visual area V2 plays an important role in converting short-term memory to long-term memory and that flaws in visual memory and deficits in object recognition may result from the manipulation of the V2 visual area.

As soon as the information arrives to be organized — before the image from the cortical areas responsible for visual memory reaches the visual cortex for processing —, the Lateral Geniculate Nucleus acts on the transfer in optical channels. During this two-way feedback process, the image is replaced and delivered to the brain's specialized cortical pathways to process visual components, a resulting signal demodulated at 40 Hz that interferes with the natural signal. Please bear in mind that in all the circuit, the ventral flow has a preponderant role in the processing, collection and evocation of visual memory, which is also affected by these intrusive images. The synchronization of brain activity in the 40 Hz frequency range serves to unify the properties represented by the neurons involved, which would only be compatible in a state of consciousness. When D2K is activated, the frequency trigger normalizes — which is the same way as when we're awake.

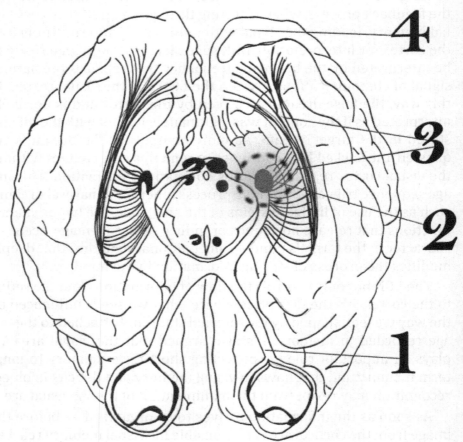

Figure 2.18 *Exact location of the Lateral Geniculate Nucleus (LGN) in the human brain.*

1) Optic Nerve – this nerve captures information through the cones and rods present in the retina that are stimulated by the light projected on objects. The function of the optic nerve is to send the luminous image that is converted into electrical nerve impulses to the brain, a structure that performs the processing of information and its storage.

2) **LGN or Lateral Geniculate Nucleus** – receives the axons of the optic nerve and transmits them to the visual cortex. At this point, images can be extracted or inserted remotely via electromagnetic transmission.

3) Optical Radiation – known as the visual pathway, which disperses and penetrates the cerebral hemispheres – occipital lobes, visual cortex.

4) Visual Cortex – the primary visual cortex constitutes the first level of cortical processing of visual information.

Conclusion

Natural dreams alone are surrounded by mysteries, profound philosophy and complex theories that have touched people's minds since time immemorial. Since then, several advanced studies have been conducted to understand these mysteries. I could write a book of hundreds of pages just about D2K and its unique ability to generate astonishing dreams that alter the target's brain until the end of their existence. However, we have to continue discussing the technology, bearing in mind that now the introspective sense, the "self" in dreams, your ego and you are no longer the only eyewitnesses to the act of dreaming. Now many people, an entire audience, can see your reactions within dreams and conduct them as they see fit at the time. Operators use the brain of others as a kind of biological platform to run any type of scenario and situation with terrible consequences for the dreamer. For this reason, once again, people attacked by this weapon feel extremely violated, humiliated and helpless.

In addition to all the problems already mentioned, turning off the mechanism responsible for blocking memories, thus forcing the target to remember perfectly the complete content of dreams, leads to serious disturbances in and out of sleep, impairing the brain's freedom in managing memory, as it alters the "inactive" period of the brain that assimilates the episodes of the day. These remote dreams are capable of directly interfering with sympathetic homeostasis and may cause other symptoms of schizophrenia over time.

This technology, if used with the person's permission and in a controlled environment, could be a powerful tool to study sleep and dream in a healthy way. However, the module is extremely harmful and becomes one of the prevailing factors for personality changes and neural reprogramming when used as an uncontrolled torture tool that replaces each stage of sleep without the consent of the victims (or human guinea pigs). Time isn't an issue here. Operators are in this scheme served by the most modern electronic devices — an advantage of supreme anonymity —, motivation and enough money to carry out this process for at least five years. This time is sufficient to conduct modern experiments, as we're going to see in the following chapters.

Every day — every moment — that we enter REM sleep, the mental degradation and confusion over memories will be presented. They thus permeate the consciousness, both concrete and reflected in dreams, destroying the mind of those who process this data in an irreversible way.

Figure 2.19 *"He woke up paralyzed and with an apparently extinguished spirit!"*

"Dämmert, Gelähmt und Mit Scheinbar Erloschenem Geist!"

— Urufaust.

CHAPTER 2.6
DANGER IN THE USE OF THE TECHNOLOGY (PART 1) - A BOY CALLED JAMES

After being aware of the existence of these weapons and how they affect human cognition, I began to see some specific events in a different light; I started to interpret certain information more carefully and notice some patterns that are inherent in this technology and in modern experiments around the world. That was when, by chance, I came across this particular story. I was able to glimpse yet another real case that illustrates, in an assertive manner, how such experiments masked in popular beliefs, tales, fantasy and religious stories, and unproven theories on metaphysical or magical subjects are conducted.

A story of the reality of experiments with electromagnetic weapons conducted across the planet on innocent civilians is going to be told. Generally, human subjects don't realize what is really happening. Dream modification technology (D2K) alone is already capable of inflicting enormous damage on the Targeted Individual, especially on children. The operators of the MKULTRA 2.0 technology (Volume 2, chapter 5) perform long-term experiments on anyone who is of interest to directly reach a specific target or individuals around them indirectly.

Operators are patient. The experiment cycle usually revolves around 5 years, but it can extend up to 10. This case illustrates the power of behavioral modification, personality modification and insertion of false memories or memories of a real story. Such memories are inserted in the target in order to transfer an experience that didn't occur with the person in real life, but with someone else. They then become false memories created using the Synthetic Electronic Dream.

We're going to see in practice what this technology is capable of doing to a growing human being and the cognitive and physiological transformations that come as a consequence of the act of altering dreams in REM sleep and instilling false long-term memories, together with the ignorance and the unprecedented nature of anomalous events.

The story portrayed — a very well-known one — is that of a boy named James Leininger. He claims to have memories of past lives of a pilot who died in World War II that were transmitted to him through terrible nightmares, as if they were real elements of his past life. His life story even became a theme for books. One of them deals with the "official" story, however, the interpretation of what happened takes on an air of mysticism and fantasy.

Before I start telling his story, I have to ask you to ignore superstitions without any proof, just as these events have been cataloged. This boy's case is the typical, premeditated long-term test conducted with a well-defined script. It achieves goals, adapts and adds content according to the child's chronological growth and biological development, with natural reflexes arising from the analysis of the data captured by the operators and the objectives achieved (or not) during the experiment, using a single weapon: D2K. Within MKTECH's range of weapons, this is the most powerful one, as I made clear in the previous chapter.

"Soul Survivor: The Reincarnation of a World War II Fighter Pilot" is a bestseller in the USA and tells how the boy began to have dreams about war at the age of two. I summarized the story by assuming that this phenomenon wasn't "reincarnation", which lacks evidence and has no scientific basis for its existence. Therefore, this phenomenon has been totally disregarded according to my understanding. For those who are curious and want to delve deeply into the details of this episode, check out the aforementioned book.

More than sixty years ago, a twenty-one-year-old U.S. Navy fighter pilot was shot down by Japanese artillery while carrying out a mission over the Pacific. Like many other pilots killed in the course of their duties, he might have been overlooked had it not been for a six-year-old boy named James Leininger. It's hard enough to remember anything before the age of 10, but this remarkable kid remembers "his past life" in shocking detail!

Parents are often very concerned when their children have nightmares. Eventually, after they comfort their children and take away their fears, the kids close their eyes and go back to sleep. Things return to normal and the bad dream is forgotten. However, when a six-year-old child, James Leininger de Lafayette, starts having nightmares at the age of four, his parents, Bruce and Andréa Leininger, were troubled. **The nightmares occurred up to four times a week, with James screaming and violently kicking upwards on his bed.** It looked like he was struggling with something or **was buried in a box**, trying to get out. The only way he could escape the nightmares was for his parents to shake him awake. **The bad dreams were out of control.**

When he was three years old, something changed. **James started having terrible nightmares about the war.** His mother woke him up while he was shouting things like: "Airplane crash! Plane on fire! Little man can't get out!".

James only watched children's TV shows, and his parents didn't remember watching documentaries about World War II with him, or even talking about the military. However, in a video of James at the age of three, it's possible to see him analyzing a plane as if he were doing a pre-flight check.

James started to be really specific with the details of his accident (inside the dreams). From July to September 2000, James began to tell his parents that the plane in his nightmares was shot down by the Japanese after it had taken off from a ship on the water. When the child was asked if he knew who the pilot was, he simply replied: James.

James named his *G.I. Joes* Billie, Leon and Walter, and said that they were waiting for him when he went to heaven. His parents later noticed that the hair color of each *G.I. Joe* matched the hair color of the deceased airmen. Billie Peeler had dark hair, just like James's doll (Billie) did. Leon Connor had blond hair, and Walter Devlin had reddish hair, such as their *G.I. Joes* versions. James said that his boat was called Natoma and remembered the name Jack Larsen.

As he flipped through an old book, he pointed to a photo of Iwo Jima in the Pacific, and said that was the place where his plane was shot down. Mr. Leininger learned that only one pilot died during the battle of Iwo Jima: James M. Huston Jr., 21. He was shot down on March 3, 1945, while on his 50th mission, before he was due to go home.

Every detail of James's dreams had been verified by the Leininger family, whether through witness accounts, personal interviews, or military records. Bruce and Andrea said that they were absolutely convinced that Huston's "spirit" touched James. They just couldn't figure out why or how.

Another time, his mother bought him a toy plane, and pointed out to what appeared to be a bomb on its underside. James corrected her, saying it was a drop tank.

When James's violent nightmares worsened, occurring three to four times a week, James's grandmother suggested the therapist Carol Bowman, who dedicated her life to the study of the phenomenon of "reincarnation". His parents took Bowman's advice and began to encourage James to share his memories. The nightmares became less frequent at once. The boy was also eloquent when he spoke about his apparent past.

Over time, James began to reveal extraordinary details about the life of the former fighter pilot. **This happened mostly at bedtime, when he was drowsy**. That was when he told them that his plane, a Corsair that had its tires deflated, had been hit by the Japanese. Interestingly, aviation historians and pilots agree that these types of tires used to suffer a lot of wear and tear. But that's a fact that could easily be found in books or on television.

James even told his father that he had taken off from a boat called Natoma, and that he had flown a few times with someone named Jack Larson. After some research, Bruce — James's father — discovered that Natoma and Jack Larson were real. The Natoma Bay was a small aircraft carrier in the Pacific, and Larson lived in Arkansas.

Bruce became obsessed. He searched the internet, checked military records, and interviewed men who served aboard the Natoma Bay. James said that his plane was shot down at Iwo Jima, as it was hit directly on the right engine. Bruce soon discovered that the only pilot from the squadron killed at Iwo Jima was named James M. Huston Jr.

James's father then came to believe that his son was the "reincarnation" of James M. Huston Jr., and that he had returned because he had some unfinished business. The Leininger family decided to write a letter to Huston's sister, Anne Barron, telling the boy's story. And in the face of so many details that he couldn't possibly know, she also began to believe in him. Unfortunately, James's vivid recollections are starting to fade as he gets older. Even so, he'll keep two precious things for the rest of his life: a bust of George Washington and a model of a Corsair aircraft. They were among the personal effects of James Huston during the war in which he died.

What really happened to the boy?

The details of the story kept coming over the years; supposedly only the dead soldier himself could provide that information. However, this isn't true. Electronic dreams transmitted during the REM phase remain vivid upon waking — they're often more vivid than everyday memories (when awake). In the mind of a child in the early stages of life, it can be aggravated. Such memories may become as extreme as those of abused children, for example. Obviously, the general population is unaware of the technology that affects the electrochemical mechanics of the brain and creates dreams of any content.

In my opinion, what actually happened was the carrying out of an experiment involving a child using only the Synthetic Electronic Dream (D2K). The capacity of cognitive modification of this weapon is already devastating for an adult in a long-term experiment. A two-year-old child, in turn — who grew up being violently *flogged*, night after night, by nightmares created in studio and transmitted by the operators behind the technology until his 8 years old — will suffer even more. His character will be deeply affected in the long run. The events that took place in a reality that was uniquely forged in the boy's mind are incorporated into the person's essence and will help to compose his personality in the following years, that is, in the next stages of development until adulthood.

This is typical of the modus operandi of the operators behind MKTECH technology. They totally changed the life of the victim and the people around him during the experiments. Well, the boy now wants to be a pilot, a military man, doesn't he? So, was this a test for voluntary recruitment? Or behavioral modification to join certain groups? It's not known for sure all the objectives of this experiment, but it's clear that the elements collected from the impact on the cognition of a developing mind were of great value, as well as the metadata and data concerning the modification of the human mind, and the returns of complex brain adaptations that developed in response to the conditions and challenges faced, including the danger of death, strong emotions associated with false memories based on facts.

CHAPTER 2.7

V2K - SOUND WITHIN ANOTHER SOUND

"I need to get off this plane, I can barely hear my own thoughts. I'm going crazy! I can hear screams that seem to come from outside the plane, calls for help and laughter. They're so loud that I am going deaf. Why doesn't anyone else hear it?"

Anonymous Targeted Individual whose brain is connected to MKTECH and is suffering from the torture of the modern V2K during a 4-hour plane trip across the country.

Resonance, reverberation and the fearsome sound within another sound, sound within noise, or voice inside murmurs.

Pure terror is the perfect definition of this modern attack using electromagnetic weapons from the MKTECH system, which has been targeting victims and causing irreversible damage to the individuals deliberately and systematically attacked by this deadly neural weapon.

The weapon known as V2K, created between 1960 and 1970, has been perfected by scientists from institutes and government agencies in absolute secrecy — and classified as Top Secret. From its inception to the present day, it has received a military-grade upgrade regarding the way it interacts with the human mind. The new generation of this weapon is as good as the original V2K, but there are great improvements in the sharpness of the sound that is interpreted by the auditory system — the positional effects that are perceived by the target —, in addition to the most maddening feature that is "to blend" with the background noise much more clearly than its previous version. It creates a complex phenomenon, capable of stunning the target and us all again! The microwave voice uses all the inherent characteristics of sound waves to create a surreal effect of "intertwining" with them. Thus, it assumes various properties of background noise, such as energy and intensity (decibels 36) and acoustic effects (e.g., echo, reverberation, distortion, and so on).

This new effect derived from the microwave sound has the main characteristic of making victims hear voices "inside" any murmur that has been picked up by the auditory system, creating a unique sensation in the individuals' brains: **that of hearing microwave voices "combine" with ambient noise. Sound within another sound, or voices within murmurs or sound within noises.** This is a very complex concept that required thousands of hours of study and practice in systematic, empirical observation of the phenomenon in progress. Any noise that is picked up is capable of serving as a "carrier" for the microwave voice, but some noises create this effect better than others. The ones that work best are sources of tonal or cyclic noise generators, such as engines, airplane turbines, air conditioning, exhaust fans, microwave ovens, distorted guitars, fans in general, powerhouse, blenders and even continuous water noise. A Targeted Individual that has already gone through this process will immediately identify that this is an attack using V2K (Voice to Skull). The target affected by this attack feels something similar to that of a person screaming in front of a pedestal fan: the result of the voice reverberating with the acoustic characteristics of the fan. However, this effect used by torture operators is much more complex than this child's play. Along with the Synthetic Electronic Telepathy (SYNTELE), V2K has become the perfect weapon for torture and remote murder.

---

36 - A decibel (dB) is a logarithmic unit that indicates the ratio of a physical quantity (often power or intensity) relative to a specified or implicit reference level. It's equal to ten times the logarithm to the base ten of the ratio of two amounts of power. The unit used to describe sound wave intensity is the bel (B), named after the Scottish-born Alexander Graham Bell, inventor of the telephone. But the unit used is actually a fraction of a bel, defined as decibel (dB).

Now I'm going to explain to anyone who has never experienced the phenomenon — and I hope they never will! — how this whole apparatus works, in addition to describing the result of the closest effect that the target feels or suffers when attacked and its consequences.

In the chapter on V2K, a lot has already been explained about hearing, the process of converting sound waves into electric impulses (transduction) and auditory reflexes. However, it's worth remembering that sound is the result of the vibrations of elastic bodies. When they occur at a certain frequency, they propagate through an impulse caused to the medium, either gaseous, liquid or solid, around the sound body, which cause transient deformations that move according to the pressure wave created. Sound is a mechanical, three-dimensional (propagates in all directions) and longitudinal (the vibration of the medium is parallel to the direction the wave travels) wave. Sound waves may suffer the wave phenomena of reflection, refraction, diffraction and interference. We're in fact exposed to various sound sources on a daily basis that can affect us in a positive or negative way. Sounds of rain or calm music provide relief and a sense of rest. However, places with a lot of conversation or heavy vehicle traffic generate discomfort and stress. Sound waves play a very important role in our daily lives and have characteristics that can help us constantly. For example, to recognize a threat: the distinct sound of a car braking, alerting us to an imminent danger.

Imagine you're walking down a street. On your walk you can hear the sounds of people around you, cars, horns, conversations, children screaming, music in the background, the engine of heavy vehicles such as planes, helicopters and buses, dogs barking, birdsong, among thousands of other sounds that reach your ears simultaneously. So far so good — almost anything that can move air molecules can create a sound. This is how the brain of a regular person processes and interprets various sounds coming from different sources, directions, of different frequencies and varying intensities, as a result of the natural dynamics of most major cities of the world. The brain of an ordinary person normally perceives all these tones, and coexists and interacts with them naturally. A person connected to the MKTECH system, on the other hand, has this entire process changed in the mind, which diverts their focus and attention on any noises that reach their ears, as we're going to see below.

Now think of each sound described above and try to imagine that they all reach your ears "carrying" other artificial sounds built in and created by microwaves — usually voices — sent by operators. It seems like an almost unbearable torture that occurs constantly, since every sound stimulus that is processed by the auditory cortex comes with clear voices. This produces stress so intense that the target is no longer able to perform their daily tasks normally. A simple walk on the street causes a surreal auditory pain. Radio waves continue to hit the victim's head on a constant basis. **As soon as a sound is captured** in a natural way, such as a bus or truck engine, such sound will serve as a carrier for the V2K. Then, the microwave voice will "blend" with the sound of the truck, making the voices amplify within the auditory pathways.

The victim is confused, paranoid, and has the impression that everyone is conspiring against them. After all, a sneeze, a bark or a fleeting sound, "carry" voices full of prefabricated studio effects before being modulated and sent by microwave pulse, acquiring the acoustic properties from the environment around the victim, such as echo and reverberation. The perception that this is a legitimate sound — a mechanical wave capable of deceiving the target — is then created. The targets think they're being pursued by operators everywhere, which motivates a false condition of omnipotence and omniscience on the part of the invaders. The victim comes to believe that they're surrounded by people who want to harm them.

A very common example: as soon as a bus passes by the target, they have a perfect and clear sensation of hearing voices synchronized with the noises that accompany that bus. Thus, they strongly believe that someone shouted from inside the vehicle or through the window. The most impressive thing is that such voices clearly follow the displacement of the sound source in relation to the target, the Doppler Effect [37], which systematically distorts the perception of re-

---

[37] - Doppler effect is a physical phenomenon observed in waves when emitted or reflected by an object that is moving in relation to the observer. This effect is perceived, for example, when listening to an ambulance siren that is moving at high speed. The observer perceives that the tone, compared

ality and deceives the brain. These phenomena occur more accurately in dog barking sounds, sirens in general, horns, motorcycle engine, noises with tonal components, pure sounds (buzzing), etc. If the target has a computer at home with a powerful cooling system to cool down the components that emit noise, it's possible to hear people's voices "reverberating", but only where that noise is being created, in addition to hearing the voices solely in that specific location and direction. It's easy to observe the "interpolation" between the fan noise and the demodulated content of the Intracranial Voice (V2K).

The closest to the auditory sensory result caused by V2K and felt by the target can be illustrated with a resource widely used in modern music, cinema and games: *Vocoder*. The *Vocoder* is very common in famous radio stations that play Hip hop music, Pop music and Electronic music, for example. *Vocoder* (a contraction of *voice* and *encoder*) means to encode, to synthesize the human voice. For those who don't know, this is a certain type of audio effect that is capable of combining with the human voice. It works like a vocal encoder. Its main function is to turn the human voice into a synthesized, robotic, grainy voice, capable of simulating monsters, extraterrestrial beings and androids. It's widely used for characters from sci-fi movies that take place, for instance, "in a galaxy far, far away". Vocoder has the carrier sound wave, the voice that mixes with that wave.

---

to the emitted frequency, is higher as the ambulance approaches, it's identical at the instant of passing by, and it's lower when the ambulance starts to move away.

Vocoder has the carrier sound wave, the voice that mixes with that wave.

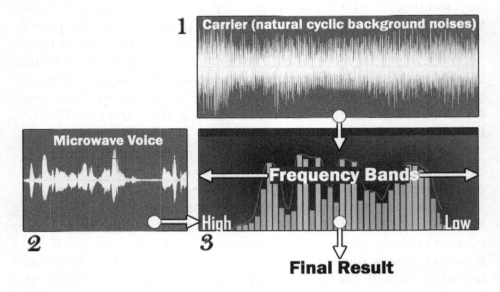

Figure 2.20 *Vocoder in action: the synthesized or robotic voice has characteristics similar to the sound of a microwave inside another sound. The singer's voice mixes with the carrier sound, giving that well-known robotic voice effect.*

1) This is the carrier sound, the noise that will merge with the voice. In the case of the V2K, these are the natural or artificial tonal cyclic noises captured by the target's ears, such as engine noise, turbines and others.

2) The voice of the singer that will be modulated along with the carrier. In V2K, this is the equivalent of the voice coming from electromagnetic transmissions that would be demodulated by the target's brain.

3) It's the final effect, the audible result in several frequency bands in which the connection between the two is noticeable; the carrier noise with the voice. **The final effect is similar in V2K, but it occurs within the target's brain.**

So, we can make an analogy between *Vocoder* and V2K. Take, for example, a Targeted Individual walking on the sidewalk and one of those old trucks. The truck goes by, emitting black smoke and noises at very high decibels, which is very common in the streets of Brazil. As soon as the sound wave from the truck engine is interpreted by the auditory cortex, such sound wave will be the equivalent of the *Vocoder* carrier wave — the background noise (figure 2.20, number 1). The V2K would be the equivalent of the modulator containing the content that will be merged with the carrier (figure 2.20, number 2) and the frequency bands would be the final result of the experience for the victim (figure 2.20, number 3), the content of the microwaves already mixed with the carrier.

The frequency waves follow the characteristics of the human voice. Consonants behave like noise while vowels behave like sounds in which there is always a dominant frequency component. In this way, some elongated words or screams reverberate or interact more clearly than others. The *Vocoder* effect is created in the studio and its product is executed via mechanical sound waves. Using V2K, the effect occurs in real time, in the target's cortex and in a very dynamic way. Thus, any noise becomes a "carrier" for the fearsome voice within noises.

An important aspect that takes place in the final result of the V2K is the amplification of the microwave voices according to the intensity of the sound energy of the carrier. That is, the voices seem as loud as the noise of the truck's engine that has just went by. Remember that sound and noise are two different things. Sounds are pleasant sensations to our ear, formed by a fundamental note accompanied by a limited number of harmonic notes. Noises, on the other hand, cause less pleasant sensations. The greater the number of notes it composes, the higher its frequency. Wherever the victim is, they will hear that kind of sound within another sound, or sound within noises, every millisecond of their life. This has catastrophic effects on an individual's brain and behavior.

The real danger arises when the person is confined for hours in a very noisy place with other people around, but without the possibility of getting rid of the constant source of deafening noise (e.g., an airplane, during air travel). Throughout the flight, the noise produced by the turbines serves as a carrier for the voices sent via V2K. The louder the noise, the clearer the voices are generated. This torture is literally maddening. The screams and voices of the MKTECH transmission "mixed" with the noise are so loud within the neural network connected to the hearing and the auditory cortex of the victim that they cannot hear their own internal voice.

The energy variation of the acoustic stimulus is so intense and the resulting sensory quality is so vivid that it's impossible to even think normally or hear other sounds in the environment. This can lead to a clinical state known as **neural claustrophobia**, leading most targets to despair, which hinders rational thinking and the ability to hear and process concrete reality and internal organic sensations.

The 100 dB noise from the turbines is now 100 dB of voices, screams, insults, cries, howls, whispers, uninterrupted music or whatever content the operators want to transmit. What used to be just a high (tolerable) ambient sound level that we got used to quickly — we can even sleep on a plane — ended up becoming a sonorous *inferno*, as such noises turned into screams and roars coming from different directions. The brain misinterprets them, treating them as a threat. The alert state is then triggered, preventing normal coexistence with the noise.

Intensity	Decibels	Type of Noise
Very Low	0-20	Silence to rustling of leaves or moderate wind
Low	20-40	Conversation at a low volume
Moderate	40-60	Normal conversation
High	60-80	Average factory noise or cacophony of the traffic
Very High	80-100	Piercing whistle and truck noise
Deafening	100-120	Noise from music festivals / A plane taking off

Table showing the types of noise and their decibels.

Source	Distance (meters)	Average dB level	Maximum dB level	V2K Sharpness Level
Airplane	20	110-120	20	Maximum
Generators	5	60-80	90	High
Air conditioning	5	40-50	60	High
Powerhouse	5	30-70	60	Maximum
Fans	1	30-60	80	High
Exhaust fans	1	40-70	80	Maximum
Distorted Guitar	2	80-90	101	Maximum
Car horn	10	90-95	108	Low
White Noise	0	0-20	30	Maximum

Table showing the estimated noise intensity and the V2K sharpness level in the brain using this particular source.

V2K has an effect on the human mind unlike anything a person may have experienced during their life. It's surreal and difficult to express in words. It creates unique, unprecedented negative emotions, and internal inquiries and questions about the fragility of the brain's natural defenses along with even deeper questions about the intellect and the violent way in which all cognitive processes can be completely and remotely modified.

Torture operators take advantage of this favorable situation to verbally attack in a continuous way. When faced with sources that generate background noise, the torture automatically intensifies. The V2K torture also occurs during interstate bus trips, as it interacts with the engine noise. For the victim — who is cooped up in the vehicle while the attack is amplified by the noise energy — the situation becomes unbearable. The target feels the need to get out of that place as soon as possible. The torture is further intensified if the target is seated near the engine of the bus.

The most intriguing thing is that other passengers around the target cannot hear the voices "embedded" in the engine noise. Only the victim is able to hear them, as the V2K attack is configured for a set of frequencies and parameters based on the victim's neural biometry (chapter 3). And not only that, the Synthetic Electronic Telepathy

(SYNTELE) remains constant. The voices or sounds respond to the thoughts of the victim who feels increasingly intimidated, overwhelmed and impressed by the event. The responses sent to complete the Electronic Artificial Telepathy cycle now interact via V2K.

This attack is commonly mistaken for a sonic weapon. In the recent attack on the Embassy of the United States of America in Cuba — which left many members of the US consulate with serious consequences —, the wounded individuals claimed during the investigation that something similar to acoustic weapons was used to carry out the attack. The injured people confused the effects of weapons known as LRAD [38]. However, V2K doesn't use mechanical waves, only electromagnetic ones. This type of confusion is understandable, since the effects on the auditory cortex are identical to real sounds. The details of this attack will be revealed in the next volume of this book.

The content modulated in the microwave pulse isn't processed through normal hearing pathways. It's actually the result of demodulation, which turns into a sound above 20 Hz in the region of the inner ear. It then "merges" with the murmurs coming through the natural hearing pathways after an intricate and complex system of transduction of the mechanical parts, performed by several cells through nerve impulses transmitted in the 31,500 afferent fibers of the cochlear nerve. After that, they're amplified, more specifically in the organ of Corti, which has an active electromechanical amplification mechanism responsible for the sharpness of cochlear tuning. The interference of the two sound waves on the same frequency (in phase) takes all the information to the auditory centers of the brainstem, which, after central processing, will originate the auditory reflexes, producing the sensation of hearing a sound within noises or murmurs, similar to the *Vocoder* effect.

It's at this moment that a recurring question arises: if you cover your ears with hearing protection, will you continue to hear the sound? Unfortunately, the answer is yes. Just test it! The next time

---

[38] - LRAD — "Long Range Acoustic Device" — is used to transmit harmful "dissuasive" tones and messages over long distances. LRAD devices come in various iterations that produce varying degrees of sound. They can be mounted to a vehicle or handheld. The device produces a sound that can be directed in a beam up to 30-degree wide, and the military-grade LRAD 2000X can transmit voice commands at up to 162dB up to 5.5 miles away.

you're on a plane in mid-flight, plug your ears with your fingers and see what happens. The sound is *not* only collected through the pinna (or auricle), but a good part of the energetic intensity reaches the inner ear through the bone structure and the head tissue. That is why this type of protection cannot achieve the desired effect — ear protection in the vicinity of the pinna has more effects in attenuating the sound. In an attempt to extricate themselves from the attack, the target may be tempted to listen to music on headphones, but this will actually exacerbate the intensity of the attack. Now, in addition to having the external carrier for the plane turbines "reverberating" through the skull, the victim will hear voices within the music itself. After all, the simultaneous combination of sounds of different frequencies and intensities gives tonal qualities peculiar to the sound.

So, there are many different types of sounds and frequencies with which the V2K can blend. In other words, the victim starts to hear the songs with the voices emerging from inside the instruments, emanating from the same sound source as if they were part of it. The microwave voice within the music is clearly heard. It's scary, remarkable and excruciating! Distorted guitars and other sounds like "white noise" [39] are a field day for the microwave voice to blend together. It's the ultimate torture, as the target won't be able to get rid of the voices in their head during the course of life. And this brings us to the details of the next chapter.

Important!

The interaction between V2K (the voices that are embedded in the signal via Radio/Microwaves) and the sound coming from the environment takes place in real time while the stimuli are processed by the target's brain. This interaction doesn't take place outside the human mind, as electromagnetic waves don't interact with sound waves normally. This correlation — this fusion — takes place only as the v2k and ambient sound are processed by the target's brain.

---

[39] - In signal processing, white noise is a random signal with equal intensity at different frequencies, giving it a constant power spectral density, capable of "merging" with any frequency sent via V2K. It's considered the perfect carrier for this type of attack. The white noise keeps the constant clarity of the voices, whether low or high intensity.

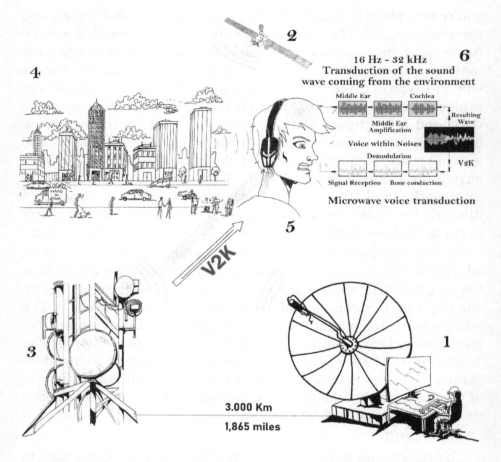

Figure 2.21 *The microwave voice "blends" with the natural sounds of the city. This phenomenon occurs due to a complex electrical interaction in the brain.*

1) Technology operators at their remote base transmit voices to satellites.

2) The satellites receive the signals from the broadcast and redirect them to antennas closest to the target, thousands of miles away from the initial transmission.

3) The antennas receive the broadcast signals and send the voices to the target's mind.

4) The natural noises of a city play the role of carrier. The noises are diverse: cars, conversations, engines, guitars, etc.

5) The target receives both stimuli constantly and simultaneously.

6) The process — in which the results of the transmissions are merged — is initiated inside the target's brain. This creates the effect of voice within noise, sound within another sound, voice within sounds, voice within noise.

CHAPTER 2.8
DANGER IN THE USE OF THE TECHNOLOGY (PART 2) - MAYDAY! MAYDAY! MAYDAY! DANGER TO CIVIL PASSENGER AVIATION

Specialized V2K torture is a type of attack that causes immediate psychological and physical problems and is overused by operators due to the auditory system's natural ability to influence and modify various areas of the nervous system, such as visual memory, creative memory, music-related memory, emotional memory, sleep regulation and level of attention, leading to common disorders resulting from excessive stress and severe brain damage. Dysfunction of brain chemicals of neural networks causes depression, anxiety, psychotic disorder, and stifles the Targeted Individual's intellect.

Microwave sounds involving words affect the target differently than meaningless noise, which activates several areas of the brain that involve language and communication. Hearing is so powerful that it outshines sight in relation to the receiver field of the message that is received from broadcast sources located 360 degrees from the individual. Small sound stimulus is enough to totally alter our physical and mental state. See how everyday sounds and noises have the capacity to exert a great influence on a person's body.

Murmurs characterize the so-called 'background noises' that are limited by the norms of several countries. They may cause the following problems:

* **Physiological problems** — in relation to the organism, they can cause deafness, nervous disorders and even death.
* **Physiological and psychological problems** — influence on the performance of human work, whether manual or intellectual.

As for high intensities, it depends on the organism. From 80 to 90 decibels depends on the regularity of timbre and frequency to affect the physiological function. However, for intense intellectual work, sounds between 4 and 50 decibels already cause a noticeable decrease in cognitive capacity and performance and produce psychological effects that are quite harmful to health.

Most bearable sound frequencies from 100 Hz to 2,000 Hz — added to the harmonious composition — directly influence cognitive health. The regularity of these everyday murmurs is bearable and goes unnoticed. However, they cause damage to the nervous system — neurological damage, mental fatigue, including damage to places in the brain directly linked to speech and hearing — when they are used as background noise for the microwave voice over long periods of time.

Let's use an example from our reality: think of that sound on TV, that music hammering in your head as you try to focus on another activity. If you no longer want to hear the television or radio, you simply turn it off, right? This possibility doesn't exist for the victims of V2K torture and SYNTELE. The sound keeps on going. So, intellectual activity is impaired, which naturally leads to anger.

And that's how one of the most dangerous aspects of this technology emerges, which is slowly taking shape and poses an enormous risk to society as a whole. It'll certainly be deadly for many people if they aren't aware of the existence of this weapon and its unique ability to affect the human brain. For instance, the attack on passengers in mid-flight. The attack on civil aviation is already taking place in Brazil as a prelude to the true intention of using this technology on certain selected people, in something larger in the near future, such as remotely-directed terrorist attacks.

Let's try to visualize the situation of a person who is about to be attacked by MKTECH and will travel on a commercial plane — a trip that cannot be postponed. Once boarding and loading are complete, and the take-off procedure has started, the target will be confined inside a plane for long periods of time with all the inherent risks and characteristics associated with flying. Commercial aircraft typically fly between 31,000 and 38,000 feet high at a constant speed of approximately 560 mph. The victim is then heading towards their destination as usual, but suddenly — midway through the trip — they begin to hear hysterical people screaming, which initially seem to come from the back of the plane. The target looks at the other passengers, and everyone stays in their seats as if they haven't heard anything. Minutes later, another terrifying and extremely "loud" scream from children or adults seems to come from the front of the plane. Once again, the individual looks ahead trying to understand what is going on. All the other passengers seem completely unaware of what is happening around them. Soon after, screams, fight simulation, confusion and voices start to shout: "The plane is going to crash! Ruuun!", "The pilot is dead", "The plane is on fire!". These screams come from all directions: from the back, the sides, the cockpit and, most amazingly, they also seem to come from outside the plane, as if someone was screaming in the distance. The effect is similar to yelling in front of a fan, as I've stressed in the previous chapter. These sounds are always accompanied by voices that interact with the thoughts of the target and among themselves.

The individual who is completely unaware of this technology will at least ask if the people around them are also hearing that sound. It'll be questions like: "Are we going to die?", "Who is screaming in the cockpit?", "Is anyone screaming outside the plane?". It could even cause a serious incident or accident. The directional, positional quality of V2K microwave sound is impressively accurate when it comes to tricking the brain with a false perception of sound direction. Keep in mind that V2K uses noise from the turbine as a carrier, so the screams will be as loud as 120 decibels of turbine noise.

This may cause problems for everyone on the plane and on the ground, especially if the target has been under previous severe torture, which is standard in most cases. The passenger may freak out, attack other passengers, try to break into the cockpit, carry out the torture operators' orders in exchange for not hearing the microwave voices or having their brain freed from the mental hijacking of this vile 21st century technology, or still blindly follow the orders as they think they're dealing with some kind of divinity, thus committing acts of terrorism that lead them to believe that they're on a divine mission. This tactic has already been used on several targets that managed, with the calm of a sage and a superhuman self-control, to resist attacks during flights and didn't go crazy. They absorbed everything in a heroic way.

The noise of the turbines mixed with the voices is an effect that causes a huge disarray in the brain and serious psychological disorders — a real terror in the heights. It's noteworthy that when individual consciousnesses come into contact, they actively act on one another — they develop within the groups, driven by the unique negative intensity of each person, which results in general panic: the so-called "herd behavior". That is, just one target is capable of ruining everyone's day, causing chaos in no time.

Imagine another scenario, much more grotesque and dangerous: a simultaneous attack on the minds of all passengers on a flight. This is a procedure already widely used by MKTECH technology operators: opening to a frequency common to all human minds in a specific place, thus generating some sounds that last for a short time and that mix with the ambient sound that everyone can hear. Now, instead of having just one "crazy" person hearing voices and causing panic, we have hundreds of different people, with different reactions, backgrounds and beliefs, different behavior and emotional reactions, who hear these horrible screams, sounds and roars. This can lead to complete chaos. There are hundreds of people in a plane, each one listening to people screaming, crying, with the feeling that this commotion is coming from outside the plane, from the cockpit, with women whispering, low and high voices with all sorts of effects that affect everyone on the plane; voices of characters from religious tales simulating demons, monsters, robots or any sound that can be created in specialized computer programs, and all this with an extremely realistic sound perception in the turbine noise power.

For a mind trained and hardened by this kind of torture, it's painful enough to endure an hour in a plane. For an "intact" mind, the catastrophic result is guaranteed. The effect is so powerful that it prevents a person from having a clear reasoning. The target cannot even hear the vocalized thought itself. It'll be really dangerous for a person who will have their mind violated by V2K for the first time. Nevertheless, it doesn't stop there. There's still another real scenario that may have already been explored by MKTECH, which among all of them is the most dangerous one: the fact that they can also use this weapon on pilots or flight attendants. There will be serious consequences if the attack is aimed directly at aircraft pilots — the plane crashing, for example. This scenario becomes possible if the most powerful and modern version of this weapon is used, such as the one used in the attack on the Embassy of the United States of America in Cuba — details in volume 2. In other words, the consequences will be disastrous. Another scenario is to attack the mind of a flight attendant who has access to the cockpit, forcing them to attack pilots or to bother passengers. One can even use people severely tortured ("Manchurian Candidates") to carry out terrorist attacks. The possibilities are as endless as the consequences.

What's more intriguing is that the Synthetic Electronic Telepathy using EMRs (Electronic Mind Reading) to hear the victim's thoughts and V2K (Voice to Skull) to send sounds straight to the cortex works without interruption in any part of the planet. Even the plane flying through storm clouds or over regions without any source of radio waves, in places without visible civilization (like in the middle of the Amazon Forest) or intercontinental oceans, the attack will still take place. The target doesn't understand how — in a remote location where there is even a "radar blind spot" — SYNTELE remains stable in the reading of thoughts and V2K attack. This is only possible with the extensive use of satellites prepared to hack the human mind, collect thoughts and direct the attack. They're known as **Neural Satellites** — secret military satellites, which will also be discussed in volume 2.

Unfortunately, a passenger airplane has no defense against electromagnetic waves in the microwave radio spectrum. The plane's Faraday cage [40], or electrostatic shielding, protects us from high charges, such as lightning, but they aren't small enough to shield other frequencies, which leaves us exposed to psychotronic weapons attacks in mid-flight. They can affect the welfare of everyone on board and on the ground, as a real danger to civil passenger aviation.

Therefore, we move on to the next case that had SYNTELE torture as a catalyst within a commercial flight, demonstrating in practice

---

[40] - "Faraday cage": a charged conductor tends to spread its charges evenly across its surface. If this conductor is a hollow sphere, for example, the charges will move to the outside surface to get as far away from each other as possible. The effects of electric fields created inside the conductor end up nullifying, thus obtaining a null electric field. The same happens when the conductor is not charged, but is in a region that has an electric field caused by an external agent. The interior is free from the action of this external field; it's shielded. This effect is known as electrostatic shielding.

the consequences and destructive power within a person's mind, capable of pushing their limits to the extreme.

I introduce you to the product of this whole obscure scheme, an important piece and representative of something that was thought impossible to happen in our reality: a remote killer. Thus, the ability to turn people into murderers without their consent is proven. These are the consequences of MKTECH's brainwashing. This recent case has obviously been distorted by the media and agencies out of ignorance or bad faith.

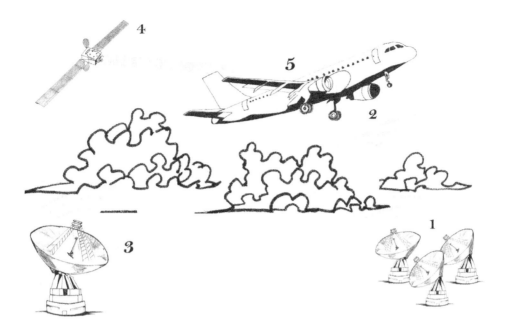

Figure 2.2.2 *Danger to civil aviation.*

1) Ground antennas aren't capable of interfering with aircraft in flight in order to maintain an endless torture; only radar and military weapons can do this.

2) Under certain circumstances, the waves aren't able to access the brains of passengers in mid-flight due to several factors, such as the curvature of the Earth and antenna capacity.

3) Then, a main ground antenna sends the signal directly to the satellite through an uplink (a link from terrestrial transmitter to satellite receiver). The satellite is capable of covering an entire continent.

4) When recruited, the satellite and its constellation, composed of several similar satellites, is responsible for amplifying the signals in the passengers' minds and for triangulating that signal with terrestrial antennas.

5) Passengers' minds are reached by mixing with the sound of the turbine via V2K and capturing thoughts via SYNTELE. It becomes an invisible, deadly weapon; a danger to civil passenger aviation.

CHAPTER 2.9

DANGER IN THE USE OF THE TECHNOLOGY (PART 3) - "WINTER SOLDIER" [41] ESTEBAN SANTIAGO

On a sunny day with few clouds in the sky — mild weather —, I decided to take a break from gathering information and researching for the book. So, I sat down next to a family member and started watching a popular cable TV news channel. Amidst the daily news, the normal programming was interrupted. There was a mysterious feeling in the air: "Breaking News!" I was sitting on a comfortable couch and drinking a cup of coffee as I watched the news in an unpretentious way. I already expected some bad news due to the sudden interruption of the usual TV show. That's when I noticed people running through the streets around an airport, several police officers with heavy weapons and equipment. What could it be? *It's definitely not good.* Then, the reporter announced that a shooting inside an US airport was taking place, more specifically in Florida.

As I analyzed that situation, I thought to myself: "Who wants to bet the shooter is under electronic torture, MKULTRA?". To my surprise, one minute later, the news reporter said that the shooter was alive and screaming that his mind was being controlled by the government. Astonished, I jumped out of my seat. The pattern was confirmed, my hypothesis could now be considered a theory: V2K torture added to artificial electronic telepathy proved unbearable for some humans inside an airplane; few are able to withstand the immense stress of a long journey with turbine noise, as we've seen in the previous chapter.

---

[41] - Bucky — a Marvel Character — doesn't remember his identity, as the Russians have reprogrammed his mind in order to turn him into an assassin known as the Winter Soldier. He was sent all across the globe, committing political assassinations with huge effects on the Cold War. However, his memory implantation caused mental instability over time. It was activated through memory triggers and he didn't remember what he did. This fictional script well illustrates the effects of mind control experiments. I use the Winter Soldier analogy as a reference for younger audiences, since this is the most popular version of the "Manchurian Candidate" from the 60s.

Esteban Santiago, 26

According to sources at ABC, the 26-year-old shooter took a flight from Anchorage, Alaska, to Minneapolis the night before, and then boarded for the airport in Fort Lauderdale, where he opened fire. A gun was checked in his luggage.

According to witnesses, Santiago took the gun from his luggage in a Terminal 2 restroom, inside the baggage claim area, and opened fire "randomly" at other passengers, reloading the revolver several times. The young man was taken into custody and didn't resist arrest, police said. According to Bryan, his brother was born in New Jersey, but moved to Puerto Rico when he was two. He grew up on the island and served as a National Guard for a few years. In 2010, Esteban was sent to Iraq, where he served for a year, Puerto Rican authorities said.

The shooter had lived in Alaska for the past few years and, according to the CBS network, served in the National Guard until August 2016, when he was discharged due to his "poor performance." Lately, the shooter was working for the security company *Signal 88*, in the city of Anchorage.

"Hallucinations"?!

In November, Santiago went to the FBI office in Anchorage and reported a series of "conspiracy theories". "I was hearing voices and my mind was being controlled by the government and the CIA," an ABC source said. The young man even claimed that he was being forced by the US government to watch videos of the extremist group known as Islamic State (IS or ISIS). Santiago's claims seemed inconsistent when interviewed by agents. Therefore, he was sent for a psychological evaluation.

The press portrayed the shooter as a mentally challenged person, but this happens in all the recent cases where people report voices and hallucinations, weird manipulated dreams. As information reaches the public and this technology comes to light, thousands of cases misdiagnosed as mental illness will be reviewed. The classic symptoms of schizophrenia are easily created by these weapons.

They use techniques to simulate the effects and thus obtain symptoms of legitimate mental illnesses. But due to this systematic torture, the target may end up acquiring the disease that was simulated earlier, as we've seen in the first chapters of the book.

In other words, the shooter became a Winter Soldier — the modern version of the Manchurian Candidate. Just picture it: your thoughts are hacked and heard, your dreams are modified every night, you suffer from sleep deprivation and start hearing voices that interact with your thoughts at all times, private cognitive processes, saturation of the afferent pathways of external stimuli with artificial electromagnetic stimuli, "auditory delusions", neural reprogramming, insertion of false memories and episodic memory problems. All of this leads to desperate acts to get out of this situation. Depending on the already highly distorted degree of perception of reality and the conscious level of loss of contact with the external world resulting from the outcome of systematic attacks of modification and neural torture, the target may commit dangerous acts without understanding the real consequences, or maybe they even think that the action performed is righteous.

Air travel is an event where operators often make it clear that they're in complete control and demonstrate that there is no place in the world that can stop them from listening and interacting with the victim's thoughts. Therefore, the target must succumb to the operators' orders — and they usually do. I would like to emphasize again that the psychotronic torture suffered in a plane is unbearable for most people. It demands coolness, calmness and unparalleled self-control to withstand its effects in the face of the noise of the turbines that charge and amplify its power to the maximum.

Not having a clear picture of what is happening regarding the massive electronic attacks using neural weapons by gangs organized as terrorist cells spread all over the world — which are engaged and involved in this type of attack — potentializes actions like Esteban Santiago's, and they become a huge problem for society. Both for the surreal ability of remote manipulation and the great damage it causes to a person. Knowing what's going on — even if it's a phenomenon hard to accept and believe — is essential for everyone right now. Perhaps, in this way, acts like this won't become a pattern, added to the everyday violence we face.

As in the case above and Alex's (volume 2), both had already reported to the FBI and local Police that they were being attacked with extremely low frequency (ELF) and microwave weapons. The case even generated a police report that wasn't taken seriously by the authorities, certainly due to the total and complete lack of knowledge on the subject. And to make matters worse, we now have the growing phenomenon of disinformation which, of course, includes these legitimate technologies of neural weapons in the "conspiracy theory" category alongside UFOs, extraterrestrials, Bigfoot, flat Earth model, ghosts, and so on.

This is one of the most compelling recent confirmed cases of how it's possible to reprogram, torture targets and force them, in various ways, to follow orders without question, as well as to modify and access all main cognitive functions and alter the core of the system that makes the human being, well, a *human being*. Access to firearms greatly facilitates the fulfillment of orders, and the post-torture insanity that ends up in disasters like these. And who were the people he shot? Were the passengers at the wrong time in the wrong place and were supposedly shot at random? Were these people, or close relatives connected to them, marked for death due to a cloudy issue? A subtle, coherent question that goes unnoticed by everyone. Would one of the "random" targets (who were shot at the airport) the object of revenge of people who didn't want to get their hands dirty? People behind the target's torture? People who paid them to kill someone without trace? All possibilities must be taken into account. These are the modern remote killers, *Winter Soldiers*, the recent version of the *Manchurian Candidates* created in the 60s.

In the end, Santiago was judged mentally competent. After all, as torture ceases, cognitive processes normalize in a short time, but their memories and suffering will never cease to exist.

A Brief Reflection

This bizarre and unusual event leads us to reflect on the culprits of this bloody episode. Whose fault is it then? The technology operators who invaded and destroyed the shooter's mind and were directly responsible for the crime, or the person who pulled the trigger? For me, they're equally guilty. I understand that this weapon is extremely powerful. It's indescribable what it can do with the functioning of the brain. People tortured by this technology succumb in a very short time. They do anything, including obeying orders to kill "indiscriminately" (as happened with Santiago) to get rid of the invasion, torture and mental parasites. He just couldn't stand the torture and the V2K influence during the flight, which was the trigger for his acts of violence, as this is indeed a unique, negative feeling. Keep in mind that the operators behind this technology are part of a very powerful decentralized worldwide organization with a multitude of available resources — both material, personal and financial —, capable of creating remote, unwitting terrorists without anyone else knowing.

Figure 2.23 Esteban Santiago shooting at Fort Lauderdale airport after MKTECH torture.

CHAPTER 2.10

EMRO - ELECTRONIC MIND READING (OPTICAL)

"I was lying on my bed, trying to read a book, but before completing the sentence, a horrible voice of a little child inside my head anticipates and finishes the sentence or reads the passage simultaneously with my inner voice. How can they see what I see? How do they know what I'm reading?"

— Anonymous Targeted Individual.

This subsystem is considered the biggest breach of privacy in history — although it's extremely complicated to choose which subsystem completely deteriorates human privacy among them all. After all, how to compare the complete loss of cognitive privacy of the EMR with the total madness reigned by the lethal V2K microwave voice? It's difficult to elect the most absurd of it all. Would it be the direct interaction with thoughts by strangers without the consent of the individual who has their mind hacked and is obliged to talk to the intruders the subject they want, at the time they want, without respecting the place, date or state of consciousness using SYNTELE? Or would it be the total loss of confidentiality in conversations and interactions with others at all times due to the EMRa — Electronic Mind Reading (auditory)? Or would it perhaps be the D2K (Synthetic Electronic Dream), the violation of the content of dreams and the outrageous reversal of sleep rhythm, which plays with the mind of an unconscious person? Although these technologies work separately, each operating in a distinct field of human cognition, their parallel processes form a powerful set that covers virtually all of the target's major mental systems, thus forming the MKTECH scheme. In fact, there is only one more module left: an important dataset for human displacement, so that one doesn't collide with any obstacles — so that we can contemplate Earth's natural beauties, such as wonderful beaches, waterfalls and landscapes. Our main sense, our sight.

Recently, flaws in operating systems embedded in devices were discovered, which opened up the possibility of activating the camera on cell phones and televisions by unauthorized persons who receive access to the interior of homes, resulting in a serious violation of privacy. We all felt frustrated and disgusted with this news! But none of this compares to the unprecedented violation of this mind-invading technology that constitutes the corruption of the human mind. The EMRo — a MKTECH subsystem — joins the other modules for a single purpose: to see what the target is seeing.

That's right! This technology has the ability to remotely access the images that are formed in the human brain, "amplifying" (reradiating) the signal that is processed by the visual cortex. Images are captured, remodulated, and sent to adjacent antennas to be decoded by advanced signal processing programs and algorithms — the equivalent of the human visual cortex, but a virtual Brain–Computer Interface (BCI). There is little data on image quality, the speed of capture of this electrical signal within the brain, and the frequency of image frames or how many frames per second (FPS) are captured. However, all targets experience the same phenomena that clearly indicate that groups are using this tool to contribute to the overwhelming sense of losing the last shred of privacy and freedom that was left in their minds.

The sight

There are more than one hundred million photoreceptors in the retina carried by 1 million axons that carry information to the visual cortex, creating visual perception, processed by neurons specialized in analyzing different attributes of the stimulus. The temporal and parietal lobe, the globules, the photoreceptor cells and all their characteristics that make up the human visual system are part of a complex integrated system that, in essence, captures the light from the environment and transforms this sensory data into images.

In general terms, it works like this: the pupil regulates light input, like the diaphragm in a camera. In brighter places, it closes, avoiding an overabundance of light. In the dark, it dilates. The range is from 2 to 8 mm in diameter — which is equivalent to magnifying up to 30 times the amount of light reaching the eye. The cornea and the lens function like the lenses in a camera, concentrating light rays sent to the retina at the back of the eye. As a lens that refracts light coming through it, it forms an inverted image of an object. In the retina, photoreceptors convert light into electrical impulses. But only one type of them, the cones, detects color. The rods, on the other hand, help us to see when there is little light. The retina uses a vitamin A derivative to absorb light at night, which is why a lack of this nutrient can lead to night blindness. Electrical impulses with the color codes, luminosity and shape (boundaries) of the observed object travel through the optic nerve to the brain. The cortex translates these impulses, perceives the movements and creates an image in our mind. The primary visual cortex, V1, is located around the calcarine fissure located in the occipital lobe. In each hemisphere, this visual area receives information from its **Lateral Geniculate Nucleus (LGN)**. The latter is primarily responsible for processing visual information received from the retina in the eye.

Therefore, we arrive once again at the sensitive region of the nervous system where information travels before it is distributed and diluted in different areas of the cortex. The information coming from the retina is analyzed by the central visual system and first passes through the thalamus, which was the target of the scientists who developed this weapon — in exactly the same way that auditory data converges to the Medial Geniculate Nucleus (MGN). The sight also has a channel where raw visual data passes from the retina and is redistributed by optical radiation to the visual cortex, which is also directly linked to visual memories. Information converges through this geniculo-cortical pathway, separated into processing channels by neurons specialized in the analysis of different attributes of the stimuli.

The **LGN** is the gateway to the visual cortex. So, for visual perception, LGN neurons receive synaptic impulses. The separation of LGN neurons in layers suggests that various types of information from vision are kept separate in this synaptic relay. Visual memories travel in a feedback way in these pathways. This massive input of impulse is where dream images, visual memories, and impulses flow straight from the retina to be decoded by the visual cortex and thus transformed into an image. That's why the thalamus and LGN hacking was fundamental to every MKTECH scheme as it was conceived, enabling it to emerge and succeed.

The **LGN** receives input from the brainstem that is related to alertness and attention. It also acts as a kind of "filter", defining the type of image that can modulate the magnitude of the responses of the LGN to visual stimuli. If there is a very explicit image that impresses the viewer, for example, the process will modulate that image with some associated strong emotion, generally creating persistent and easily accessible memories. That's why scenes of war, death and violence shock most people. Over time, of course, overexposure to certain scenes can dull the associated emotion. This same mechanism applies to visual memories. The whole state of consciousness is involved. In terms of time, the representation distances itself from the original pure sensation, it loses the importance and the positive or negative meaning of the representation of the past.

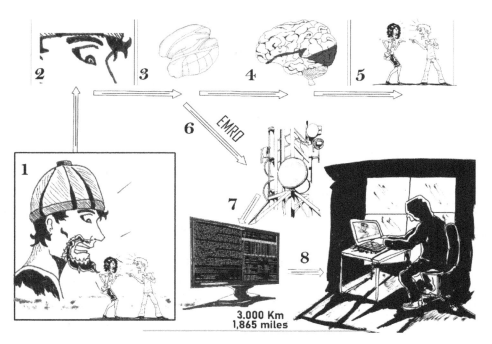

Figure 2.24 *Steps 1 to 5 represent the process of transduction of light. Steps 3 to 8 represent the capture of images straight from the target's mind.*

1) The target observes people arguing loudly in the street.

2) The eyes receive these waves in the visible spectrum, process the signal and send this information via the optic tract.

3) It reaches the LGN (lateral geniculate nucleus of the dorsal thalamus).

4) The LGN projects to the primary cortex where images will be distributed over a large area of the brain to acquire all of its visual properties — the parts of the brain highlighted in black are the key areas of sight.

5) The result of this whole process, the conscious visual perception.

6) At the moment the signal travels through LGN, another intrusive radio signal transmitted from antennas or satellites amplifies this signal.

7) Antennas capture this amplified signal and remodulate it, sending it to other antennas where it will be received, demodulated and sent to a Brain–Computer Interface (BCI) system that will interpret this signal in a similar way to what the cortex does in the human brain.

8) The operator — in their hiding place — receives these processed images from the target's brain thousands of miles away, and they successfully see what the target sees in the moment.

The resolution of captured images and the frequency of frames vary according to the group involved and the purpose of the psychotronic attack. Each military, intelligence or terrorist group uses different equipment and infrastructure, some extremely more advanced than others. It's likely that the resolution and frames per second on this high-end equipment maintain the visual fidelity of the human eye and the number of frames displayed, which is around 30 frames per second. Organized gangs use this technology that was probably "rented" from the Russians — acting in collusion with Cuban terrorists — in Brazil, and the resolution and the frames are likely not that good. However, it's known that this module is present in the weapons used by national terrorist cells and groups — the details will be explained throughout the book and, more precisely, in volume 2, chapter 11.

A target who was trying to read a book at their home was surprised by the anticipation of the text via microwave voice coming from far away, making it clear that operators could see what was in the target's field of vision in a remote way. They could even focus enough to catch the tiny letters on the pages in front of them. This incident took place every night before bed. Besides, this strange voice can also read along with the target inside their own mind. This prevents the act of reading and absorbing the knowledge contained therein.

In other different circumstances, it's possible to observe the complete mastery of the visual environment and the scenario in which the target is inserted. **This is the only MKTECH subsystem that doesn't provide immediate feedback to the brain, but it's possible to know that they see what the targets see when they use daily visual images** — that only the victim could have seen — to compose the cinematographic content of dreams. So, the target is confident that operators can see everything they're currently seeing. This is yet another torture strategy perpetrated by the cowards hiding behind the technology that drives someone crazy — in their own home!

In a controlled environment, experiments in monkeys with brain implants show this technique in a more detailed way. MKTECH uses the same principle, however, remotely, without the need for physical contact with the victim. The reading of signals in the visual cortex can be carried out in a laboratory. The experiment "*Computer records animal vision in Laboratory - UC Berkeley*" demonstrates how to perform the reading of a primate's visual cortex. With this, it's possible to see what the individual is seeing. The video is from 2009, but it demonstrates how it's possible "to see" by monitoring the visual cortex.

2.10.2 - Visual memories of thought

Violating these thalamic pathways is in fact a very ingenious technique, as it isn't necessary to trace the paths by which thoughts are formed, measuring the complex behavior of billions of neurons. All you have to do is amplify the signal and capture it ready to be interpreted by the cortex responsible — by a common channel, the THALAMUS. With this, it's possible to capture images coming from the eyes at once, that is, to capture images of visual thoughts that travel through the optical pathways, and insert and collect images in dreams during the sleep and REM state. This smart conclusion after decades of MK-ULTRA experiments brings us back to the expression "kill two birds with one stone".

Imagine what it'd be like to chase these signals into, for example, thirty different visual cortical areas situated between Brodmann's area 18 and 19 in the occipital lobe, as well as neighboring regions of the parietal and temporal lobes, rich in specific cells associated with information processing channels with high degree of specificity — whether for spatial vision, shapes, colors or action. The brain always works using analogy, deductions, abstractions and inductions; words often rely on mental images to automatically contextualize the object being thought or the word being said. Whenever we remember a fact, a person or a pet, this mental image can only actually exist if it's been seen before. For example, if I ask you: have you ever seen an animal called "Kräkän"? What mental image would be attached to that word?

There are no concrete images, just random ones in case you've never heard of it, right? Now I can tell what it really is: a giant octopus-like sea monster from Scandinavian folklore, capable of sinking ships with its suckers in a deadly embrace — sometimes as big as a blue whale. Notice how many mental images immediately make the word take on a visual meaning equivalent to reality. Every image associated with those words — like the whale you imagined — was recruited into your visual memories. Once it was triggered, it traveled through the optic tract and optical radiation, and through the LGN. As long as this feedback process is active and the signals travel through the cortex and thalamus, everything will be stolen from your mind.

Mental images are extremely important to us. With them, we imagine. They lead us to the inventive and creative process. We create products from our creativity based on what we can see. The creativity of a product lies in the extent to which it restructures our universe of understanding. Thought images compose a representation of the world in a symbolic form of images.

The uncomfortable fact that third parties have access to your mental images around the clock brings a multitude of problems for the individual and for humanity as a whole. Imagine if everything you're thinking about at this exact moment was played on a screen in a public place for everyone to see. Now imagine this adding to the fact itself, environmental and social stimuli being activated and deactivated, and stirring these mental imagery ingredients like a "cauldron of soup". The images displayed on the screen would be unpredictable, intense and somehow uncontrollable, because that's exactly what happens to the victim of this weapon.

Another feature of this subsystem is to capture the images of visual memories that are used as the basis for remote dreams via D2K. It's the dreams that cause the most confusion, that mix real images captured from the target's memory or vision, and insert them into a completely different reality from dreams, as we've seen in the chapter on artificial dreams.

What used to be just confined in people's minds — what was just a thought — is no longer just that. Any thoughts generated in their minds, in any condition and under any circumstances, are amplified, electronically extracted and stored. Thoughts can be used against you when it suits the operators, and can also be used completely out of context. This is the power of EMRo: mental/visual images and dream images are easily stolen and inserted by those who have access to the technology. They're seeing what you see right this minute.

Any electronic apparatus that interferes with the free traffic of mental images of any human — that captures everything that is being observed by them — must urgently be the subject of a wide debate by society in order to decide what measures to take regarding the theft of our mental/visual projections. One of the initial measures should be that of helping the people impacted by these experiments to seek compensation for the violation of their privacy, thus discouraging this increasingly common practice.

CHAPTER 3 - PHYSIOLOGICAL MONITORING - NEURAL BIOMETRIC SIGNATURE - REMOTE POLYGRAPH

RNM - REMOTE NEURAL MONITORING

This module is of great importance for conducting the entire MKTECH scheme, due to several unique characteristics composed of three distinct elements converging towards a purpose that is achieved in a joint effort: to monitor all possible raw physiological data as well as all electrical brain wave patterns, which indicate the individual's current physical and mental state, assisting the torture and physiological (reactive) responses that are of great value to the tests conducted by the operators.

The RNM subsystems complement each other, forming an essential module for the scheme. Thus, we have the biometrics of the brain's electrical functions, making it unique in the sea of other signatures. On Earth, there are approximately 7.5 billion of these electrical brain signatures that correspond to every living individual. This biometric data and brainwave patterns are constantly monitored by the equipment responsible for maintaining the connection with the target's brains. In addition, they ensure that any doubts concerning the information captured from the individual's thoughts are quickly clarified by going through a screening system that will indicate whether a given thought is true or false: the Remote Polygraph, the modern version of the lie detector. In this case, however, it's used to analyze thoughts before them being expressed in behavior or in physiological reactions.

This set of systems is directly responsible for monitoring each step of the target — wherever they're physically located —, collecting telemetric data based on the electrical activities of the brain, further contextualizing their current general physiological condition, as well as their emotional state, calibrating the accuracy of attacks, which makes them increasingly destructive and efficient.

3.1- Neural Remote Biometrics

The first system of fingerprint identification dates back to 1891, when Juan Vucetich started a collection of fingerprints from criminals in Argentina. Josh Ellenbogen and Nitzan Lebovic argued that Biometrics originated in the identification systems of criminal activity developed by Alphonse Bertillon (1853-1914) and by Francis Galton's theory of fingerprints and physiognomy. According to Lebovic, Galton's work "led to the application of mathematical models to fingerprints, phrenology and facial characteristics", as part of "absolute identification" and "a key to both inclusion and exclusion" of populations.

Many different aspects of human physiology, chemistry or behavior can be used for biometric authentication. The selection of a particular biometric for use in a specific application involves a weighting of several factors. They identified seven of these factors to be used when assessing the suitability of any trait for use in biometric authentication.

* **Universality** – it means that each person using a system must possess the unique usable trait, or characteristic.
* **Uniqueness** – it means that the trait must be sufficiently different for individuals in the relevant population that they can be distinguished from one another.
* **Permanence** – feature that must be immutable, relates to the way in which a trait varies over time.
* **Measurability** (collectability) – it refers to the ease of acquisition or measurement of the trait.

* **Circumvention** – it relates to the ease with which a trait might be imitated using a substitute or artifact.
* **Performance** – it refers to the accuracy, speed and robustness of the technology used.
* **Acceptability** – it relates to how well individuals in the relevant population accept the technology in such a way that they are willing to have their biometric trait captured and assessed.

Biometric characteristics can be divided into: physiological and behavioral. Physiological biometrics involves a person's physical characteristics — for example, fingerprint, face recognition, iris recognition, palm print, hand geometry, DNA, voice recognition, electrical activities of the brain and the anatomical shape of the mouth. Behavioral biometrics can be the person's handwritten signature or another mechanical motor activity that is unique. For instance, recording parameters of the person's walking along a route for a certain period, creating a kinematic model with motion capture programs and so check distinctive details such as stride, posture, movement of arms and shoulders.

For the development of psychotronic weapons and the MKTECH itself, models based on the analysis of electrical activities carried out through the remote EEG (Electroencephalogram) are used and mapped. For this purpose, several mathematical models and algorithms were created by scientists involved in the evolution of this weapon to individualize the brain waves mapped more and more precisely. The BCI is responsible for receiving this information in real time, quantifying it appropriately and forwarding it to another responsible algorithm or subsystem, translating it into a unique pattern for each individual on the planet.

The BCI is able to receive the data and calculate the magnitude of a set of neuronal firings, using different stimuli. These stimuli can be external, such as a response to a visual, auditory or both stimuli — an electromagnetic feedback stimulus based on the set of momentary electrical activities or thought images. Furthermore, as MKTECH doesn't only work with the raw electrical signals, but with the demodulated content of these signals, objective and subjective constants and variables can be added, improving the individualized margin, manipulating particular characteristics such as the way the target focuses on certain contents, activities in which they use certain areas of the brain, parameterizing the EEG signal that is conditioned by mental activities performed at a certain moment in time. Neural biometrics thus has a direct dependence on electrical signals, and a behavior influenced by the time interval in which this configuration is acquired.

It's not trivial to create efficient neural biometrics. We go through a series of problems, e.g., dealing with the principle of firing of neural populations, reduction in compensatory metabolic responses and variable firing rates depending on a series of factors (level of consciousness, intensity and type of sensory stimulus, for instance). Even the same stimulus can generate different electrical patterns. Thus, the P300 appears as an EEG signal peak. Approximately 300 milliseconds after the presentation of a relevant stimulus, the complexity of that stimulus and the individual differences in amplitude and latency — ranging from 250 to 500 ms — make the P300 essential for the studies on EEG signal decomposition, serving as a basis for others analysis and individualization programs of brainwave patterns that enable the transformation of these waves into a unique pattern, measuring electrical potentials over a certain period of time for each individual.

It's possible to evoke auditory, visual and other stimuli potentials specific to MKTECH protocols, in which one can measure the level of reaction to a given stimulus of signal feedback, stimulating their own mental reactions without the target noticing that this signal is furtively modifying certain frequencies in the brain. You can specify any location on the planet and send radio frequency waves similar to the way that radar works. Return waves are processed by a biometric system, as in the identification of any object by radar. Every time a brain pattern is identified, its signature is registered, identifying that pattern as being of a specific person.

Other similar algorithms come from a complex series of mathematical equations that can individualize the brain of every person on the planet with accuracy if compared to fingerprint, iris or palm print. They are essential for the MKTECH scheme, as these data sets will be digitized and transformed into a brain ID, an identification number, a PIN, a node, token or virtual profile, which will then be sent to the satellite through uplink transmission so that the equipment responsible for the telemetric triangulation will later know exactly the geolocation of the target. In other words, **once your Neural Remote Biometrics is mapped you'll be followed wherever you go in any part of the world, by satellites, terrestrial antennas and advanced AIs made exclusively to assist the technology of invasion, control, reading and torture of the mind (MKTECH).** The frequencies of this mapping occur between 420 Mhz to 750 Mhz, which are the resonant bands of electrical activity in the human brain.

Similar to ping [42], the system maintains constant contact with the target and tests the connection speed between them and the equipment. It can even be used to understand the distance and position of the target. Now the victim knows why torture is never "turned off", neither is the Synthetic Electronic Telepathy, nor the V2K, in mid-flight or while traveling abroad, and why other people around them cannot hear what they hear.

Just the act of building remote biometrics without your consent or that of anyone else violates all constitutional principles of privacy and freedom. We must be aware of these acts, because the violation of biometric data within the individual's home or wherever they are is the beginning, the "kick" that culminates in barbarism conducted by these weapons.

There are several ways to enable biometrics. The most common is to use electrical characteristics (brain waves) of some parts of the brain and measure them for a certain period, which forms a unique pattern for the Targeted Individual, that is, their neural binary representation. Once the unique patterns are mapped, it's possible to establish the set of frequencies that will interact with the target's brain, delimiting the frequencies in the radio or microwave spectrum where demodulation will only occur in that specific brain. Thus, this target will exclusively be able to interact with the content sent over the waves in a set of personalized frequencies using

---

[42] - Ping is a computer network administration software utility used to test the reachability of a host on an Internet Protocol network. It's available for practically all operating systems. It operates by sending packages to the destination equipment and "listening" to the responses. If the target equipment is active, a "response"— the "pong", an analogy to the famous ping-pong game — is sent back to the requesting computer.

SYNTELE, V2K and D2K. Other brains won't undergo this interaction — they won't demodulate the wave content. Some remnants or natural demodulation can be felt by other brains, but never in the intensity to which the target in question is subjected. The Neural Remote Biometrics will tell the systems which frequency range each mapped brain area of the victim is operating via Remote EEG, indicating to the system which set of frequencies to use.

Therefore, we can draw a parallel between common radio stations and the individual's mind. One could say that after having the signals captured, analyzed and classified by an individualization algorithm, the result of this process will be the equivalent of measuring a fixed radio station frequency: 88.3 Mhz: Target A radio, 90.3 Mhz: Target B radio, 99.2 Mhz: Target C radio. In this way, the brain of the target would resemble a fixed radio receiver in which it would be possible to tune only to that specific frequency. Other frequencies in the spectrum, intended for other brains passing through this "receiving antenna" (your mind), wouldn't be demodulated, as they don't have the "tuner" in that specific range.

These electrical data can be collected as samples of several automatic mental activities, different mental tasks that have distinct output signals. Some neural interactions can be added to feedback signals of the algorithm of creation and constant verification of the biometrics.

There is also an algorithm that does the encoding — the translation of these electrical signals into digital commands — similar to the shift from analogue to digital TV broadcasting. This happens as follows: coding by sampling retains the sampled analog signal for a short time sufficient for encoding and quantizing. Decoding, converting binary codes to a corresponding voltage level, assuming 16 quantization levels corresponding to 0 to 7.5V (volts) at 0.5V intervals.

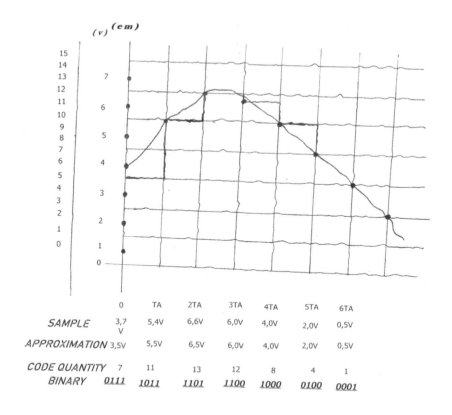

Figure 3.1 *Conceptual example of the quantization of sine waves extracted from the EEG.*

Evoked potentials can have various origins in the type of stimulation. So, we have, for instance, the visual evoked potentials — resulting from the presentation of visual stimuli —, and the auditory — resulting from the presentation of sound stimuli to the subject. It's also possible to use a standard signal of both a visual and auditory stimulus. By using electrodes placed on the scalp, it's usually possible to obtain several channels coming from different parts of the brain, such as the visual and auditory areas. One can monitor more than 50 different channels from various areas, and mix with the data derived from the general state of the target — calm, stressed, attentive — given the variation in the way each algorithm works within its degree of complexity. After that, an algorithm is used to create a checksum identifying the individual. This checksum is produced from different areas of the brain, which have different variations in individuals. Hence, the individual's brain frequency is represented as an identifying number of the electromagnetic structure of that brain. Or, to put it simply, every brain in the world can be mapped, constantly scanned and located by satellite anytime, anywhere!

The analysis of EEG signal processing and decomposition is an extremely vast field and contains numerous equations and advanced technical terms. I've tried to summarize it as much as possible for a better understanding of this subsystem in the MKTECH context.

3.2 - Telemetric EEG, Electronic Brain Link, Remote Neural Monitoring

This is a vital module for the functioning of the MKTECH system, responsible for all data referring to the target's raw brain waves and configurations that indicate various aspects, including the current emotional and physiological state and the position of the body in space in real time. It's possible to create a virtual avatar that displays the body position in space in real time of all victim's limbs in a graphical representation with the electrical feedback data obtained. Thus, they visualize in which position the body is in a defined period of time. They remotely monitor brain waves and electrical changes, biomagnetic analysis system similar to ECG, EEG, EOG, EMG, EGG, Respiration Rate, Pulse Rate, Temperature, Impedance Cardiography and Electrodermal Activity. These data can be captured or inferred with a satisfactory margin of approximation if compared to the equivalent of using electrodes, as well as brain functions and firing patterns of neurons involved in certain behaviors over a period of time, metabolic activities, among others.

The term EEG is used here only by naming convention. After all, it's possible to measure other elective data that aren't part of the EEG universe. This module is capable of measuring many electrical, magnetic and physiological parameters, thus providing a complete picture of the Targeted Individual for decision-making by technology operators and support for other technologies.

EEG is based on a mean of a synaptic current of an action potential produced by thousands of cortical neurons that carry little neural information. Electroencephalography is defined as the measurement of electrical activity produced by the brain. Neurons generate action potentials that have an electrical aspect, and, therefore, the set of activity of the neuronal networks that make up the brain produces a range of electrical potentials that can be measured under the surface of the scalp or directly on the brain via electromagnetic waves. The measurement of such potentials doesn't focus on the activity of a single neuron, but rather on the combined activity of millions of them, as individual cortical neurons and their probabilistic firing patterns can participate simultaneously in multiple neural populations. This signal acquisition results in the Electroencephalogram — EEG, which is an increasingly important means for technology operators given all the information that can be synthesized from it in a few seconds.

Telemetric EEG primarily measures the voltage produced by currents that flow during synaptic excitation of the dendrites of many pyramidal neurons in the cerebral cortex. Furthermore, it allows a series of mental functions to be read, monitored and altered. Contrary to popular belief, the reading of brain patterns by EEG can be done without electrodes. They're actually used only for convenience, ignorance or commercial interest. EEG is then the sum of synchronized waves from certain mapped areas of the brain. Keep in mind that this single area can be studied in isolation and is also used as a gateway to hack into certain areas of the cortex and violate the content of thoughts.

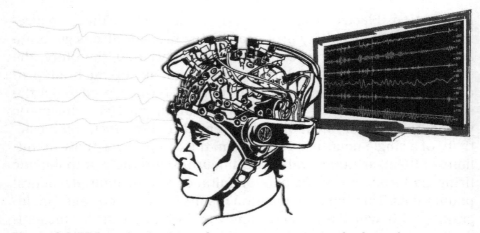

Figure 3.2 *EEG monitoring in synchronous waves using scalp electrodes.*

The powerful AI (Artificial Intelligence) of the remote EEG module by telemetry acquires the electrical particularities of the target — the personal settings of the neural waves — through constant monitoring in everyday life, checking the normal neural state that occurs on a daily basis in the face of social interactions and interactions with the environment: worry, anger, relaxation, attention and focus on a certain subject, sadness, boredom, and so on. The target is observed 24/7 in their personal life. The way they interact with the normal, everyday circumstances inherent in almost every human being living in a modern society is electrically observed. Interactions such as: relationship with a certain group of people, or a particular person, their anxieties and fears about a certain event and life issues that afflict most people, for example, work, money, family, education, and of course, the deeper problems of the human psyche. Virtually all patterns of physical, mental and emotional states are captured, producing a catalogue of abstract emotions, which are mapped on individuals in conjunction with the configuration of this set of waves assembled according to each state reached in this catalogue. Many of the parameters are received using electronic torture reactions via Synthetic Electronic Telepathy. Keep in mind that the representation of sensory events doesn't originate from isolated individual neurons, but rather as a circuit that extends from the peripheral sensory input.

By checking which stimuli cause well-being, which are neutral and which cause the opposite feeling, they constantly monitor all EEG settings, forming a picture of various parts of the brain in different everyday situations. The AI responsible for learning to interpret these emotional states based on EEG data inputs and the set of wave patterns quickly adapts to the target's personality. The AI is able to indicate to operators and other AIs in the system precise information about the cause-and-effect relationship between the target and the environment, conditions responsible for their behavior. If affected, they're classified into certain attributes and internal variables along with the small electrical changes and nuances in personality and character variations.

The psychological state of someone who perceives an external event is a determining factor for emotions and their stimuli. Perhaps the AI, after a certain period of work, using only these observations, creates a virtual representation that perfectly indicates such states. It gets to know the target so well due to their responses to stimuli in certain circumstances that it can even exceed the knowledge of a person who lives together with the target, such as their mother, spouse or siblings. This accurate remote monitoring carried out against the target's will in their intimacy/privacy ends up generating thoughts that reflect on various aspects of their existence, which takes a great deal of time. Reflex thoughts following feedback become part of the mass of data that indicates to psychotronic torture operators the victim's current state of mind. The intention is to stimulate an aggressive and distressing state, blocking attention and diverting the victim's normal flow of thoughts. So, operators use this data to constantly modify these points.

When the target's state is relaxed or with an unwanted focus by operators (usually linked to well-being), which leads the person to internally ignore attacks on vocalized thoughts for a brief period, the AI automatically initiates the process of changing the state. It then increases the intensity of the attack on the brain and chooses a subject that cause an organic reaction in the target, such as anger. This predetermined weakness is systematically attacked (via V2K). The target's state will be invariably altered, so their attention focuses on the attacks and the cognitive domain by the operators returns. The inevitable change occurs both by the unprecedented invasion of privacy and by getting into matters of their private life that the targets don't want to think about at the moment or don't want to share with anyone else. After all, the decoding of the meaning of the words — in all their complexity — invariably alters psychophysiological states.

The TELEMETRIC EEG serves to map emotions and the person's momentary emotional state. Emotions are externalized; a behavioral expression that manifests through a complex series of responses from autonomic and glandular systems. Unlike the others, this module deals only with raw data, the electrical analysis to be formed within a machine learning neural network of a certain pattern capable of recognizing accurate emotional states and predicting possible subsequent behavioral reactions to such states. Remember that the target is being constantly bombarded by stressful MKTECH stimuli, as we've seen in previous chapters. Another functionality of this module is to indicate the target's body position in space in the form of a 3D avatar based on telemetric data received via remote EEG.

3.2.1 - Body position in space

The target's frame can be transformed into a reliable virtual representation that will update their position in space according to their actual position inside their own home, for example. This technology is known as *Through-the-wall surveillance* (TWS). The victim is shocked to discover that when they walk through their house, their every step is being monitored. This technology is used in conjunction with other techniques for capturing electrical signals. Remote Electromyography — which is included in the EEG — is even a device for monitoring the electrical activity of excitable membranes of muscle cells, representing action potentials triggered by the electrical voltage reading. It helps to form an image composed of many dots that show the position and location of the target inside their residence. These captured waves are sent to a program that creates a visual analogy in a 3D avatar that depicts the victim on MKTECH computers in real time. This is how it happens with everything that can be monitored in isolation — or the electrical dynamics — that form perfect, positional and physiological representations, which updates the target's position in the virtual space according to their real position in their workplace, for instance. Monitoring of peripheral somesthetic pathways to capture the position of the moving body. Facial expressions, body posture and physiological response are some ways to measure the intensity of emotional pain felt at the time, so important for conducting the experiments.

The system's AI constantly monitors the individual, assimilating behavior patterns. The state of attention and focus are even identified and compared to a series of preprogrammed patterns and learned via machine learning by artificial intelligence that controls all these attribute classifiers and determines the target's current physiological state. Thus, the system analyzes the data received, setting a physiological configuration and general state that would be ideal for imposing greater mental suffering, inattention, anger and acute stress. This analysis is then forwarded to another AI that will conduct the attack straight to the target's mind based on these parameters. Electrical details, such as: solving or ending a problem, subtle waves also enter the big picture, intercepted in EEG spikes, in an approximate pattern of behavior.

A simple example that runs behind the scenes of the program is the transformation of frequency electrical potentials into graphs. It's known that the frequency generated by the eye region separated by the analysis algorithm shows that, when someone has their eyes open, the received signal is at the frequency of 4 Hz; with eyes closed is 5 Hz. And so, signals are captured from all over the body, helping to contextualize movement and body position based on inferences such as these.

If redundant systems fail momentarily — such as the EMRO - Electronic Mind Reading (optical) or the EMRA - Electronic Mind Reading (auditory) — it's possible to deduce the activity that the target is currently engaged in by their body position. That is, if they're sitting, standing, lying on their back, face down, eyes open or closed, legs open, closed or crossed, a position similar to the act of driving, or if they're working out or performing any other activity, inside or outside their own home. It's useful to deteriorate the target's privacy by capturing details, noting the positions in which they sleep, if they're doing their physiological needs, taking a shower, and so forth. These moments are narrated by third parties and sent to the target's brain constantly, which intensifies the madness.

The coexistence that the target is forced to endure in their privacy permeates their entire inner self. This is a violation of privacy regarding their body image. The program, this kind of "body space hacking", accurately infers the activity the victim is currently engaging in, as well as the location of their limbs, which aids the AI's work. The analysis of electroencephalic changes in relation to the target's mental states that aren't in line with the primary objectives of torture aimed at mental degradation is immediately rejected and countered by MKTECH. Don't forget that one of the main intentions of SYNTELE and V2K is to keep the individual's full attention on the content of the messages sent, making directional focus unfeasible, blocking concentration through emotions produced by the content of the attack, keeping as much as possible any other everyday action hampered, like reading a book, watching a movie, studying and working. In this way, the system knows when the target is focused or not on the torture, and it directs their attention to the content delivered by operators. If the parameters related to the total focus on attack content and settings on general well-being don't conform to the expected result in the analysis, the AI — based on these parameters and relying on its acquired knowledge of forced coexistence — is able to resort to tricks to accentuate the *noise* in the mind, including increasing the volume of the attack within the brain. It's also a tactic used to fill up working memory. We'll see more about our biological memory more precisely in section 5.10, volume 2 of the book.

Feedback signal with its reference displayed in spectral maps, projections, three-dimensional graphics, neuronal response histogram around a given stimulus also support human and autonomous decision-making. The modification of attention, the change of focus and parietal reasoning in the tactile receptive field ends up affecting the bimodal cells that respond to stimuli coming from two distinct sensory modalities: adjacent extra personal space known as "peripersonal space", increasing or decreasing the ability to perceive the space around us. This is another harmful effect of this attack and its capillary force in the most diverse human systems.

The Telemetric (Remote) EEG is the main module used by all other modules of the systems. It's responsible for mapping all areas of the brain in real time, captured by radio signals from external antennas "pointed" at the target's cortex. The database that operates the remote EGG module has all known brainwave configurations previously mapped to each part of the brain separately. EEG rhythms are categorized by their frequency ranges and are often correlated with certain behavioral states, such as levels of attention and agitation. In the mammalian brain, for example, synchronous rhythmic activity is coordinated by a combination of pacemaker and collective methods generated by the thalamus. This turns the action of the other cells into a standard rhythm that can be registered on the EEG and interpreted absolutely and accurately.

Emotions that were previously only perceived by the person's description or through behavioral observations — such as gestures, body posture, movement — can now be electronically measured in a remote way, along with physiological data that constantly check for certain organic changes, the electrical conduction of the skin that increases with the individual's degree of emotional excitement, volume and some known human-inherent wave patterns captured by EEG, blood and heart pressure, change in body temperature, exudation, salivary gland secretion, muscle tension and tremors.

Operators monitor, among other waves, those responsible for indicating sexual arousal in the target. It's widely used in the Synthetic Electronic Dream (D2K) to violate the most private desires of human beings. A condition that shows to operators that the target is sexually aroused, with intentions or thoughts linked to sex, desire and the intention to carry them out, is created. The AI will detect these impulses. The attack is then configured to discourage the target from taking any action related to that matter, sending inhibitory stimuli via SYNTELE that were previously mapped, which inhibits the sex drive.

A recurring practical example comes from the auditory discouragement which, in the case of V2K, is more sophisticated and has a clearer auditory interpretation than the normal pathways. If a male target is heterosexual, stereotypical voices of homosexual men talking about sex will reach his auditory cortex in order to immediately discourage him.

3.2.2 - Brain waves

Some primary waves of states of consciousness form the initial basis of the system. After some time and with a forced coexistence with the target, the AI verifies details and small changes in the waves. It thus creates a pattern of behavior based on the emotional picture observed several times over the course of events and is able to predict with great precision the next reactive behavior in the face of certain types of psychotronic attacks. Here are the EEG waves of physiological and mental state modification:

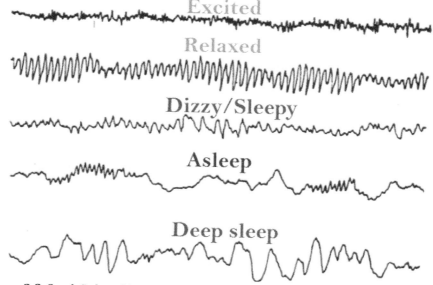

Figure 3.3 *Good: Relaxed/Focused. Bad: Restless/Nervous/Anxious.*

Motivations and intentions

AI's unrivaled ability to capture all kinds of feelings and states by interpreting subtle particularities in each Targeted Individual creates immense understanding in the personal sphere. For this, it uses neurological representations and patterns of collapse, characteristic of observation and learning, based on several factors, among the main ones are the psychological, physical and electrical ones.

Factor 1: Primary emotions:

The signal that is received with the settings containing the oscillatory rhythms can be quickly measured by providing changes in any rhythm separately, or in its entirety, to modify the target's current emotional state.

Factor 2: States of consciousness:

The technology also works by altering the brain's electrical signals at a distance, either by brute force, psychic driving or by continuous development of an extremely negative condition.

The main waves captured by the brain's electrical activities are:

Alpha[α]

Figure 3.4 *This activity corresponds to the 8-13 Hz frequency band. These rhythms can be found in most adults when they're conscious and in a relaxed state with eyes closed. They mainly manifest in the occipital area. This type of rhythm doesn't present itself in deep sleep states and is strongly attenuated in episodes of mental concentration for a particular activity that recruits great power of focus and mental processing.*

Beta [β]

Figure 3.5 *The frequency range associated with this type of rhythm is the band between 13 and 30 Hz, and is subdivided into three sub-bands: β1 (13-18 Hz), β2 (18-24 Hz) and β3 (24-30 Hz). The need to subdivide the β rhythms into these 3 sub-bands comes from their different origins. However, in general, it can be said that these rhythms are responsible for processing information from outside and for problem solving and decision making. They're thus involved in mental operations that require a greater degree of complexity.*

Delta [δ]

Figure 3.6 *They correspond to slower brain activity (below 4 Hz) and are revealed in the deep sleep state of adults (stages 3 and 4), and newborns.*

Gamma [γ]

Figure 3.7 *These rhythms are the most complex rhythms acquired in EEG. They're associated with perception and consciousness tasks and occur from 30 Hz up to the band limit determined in the acquisition of the EEG signal. They're presumably the most information-rich brain rhythms.*

Theta[θ]

Figure 3.8 *Theta rhythms occur in the frequency band between 4 and 7 Hz. Its name comes from its location (thalamus). These rhythms in adults are symptoms of brain dysfunction and are also revealed during episodes of stress, light sleep, meditation or hypnosis. They are, however, considered normal during childhood.*

The target's EEG is constantly — that is, every millisecond of their life for long periods — monitored. MKTECH's goal is to inflict as much internal damage and suffering as possible. As soon as any of these rhythms considered beneficial is detected, the program's AI analyzes and warns operators that a configuration not allowed has been identified, which assists human decision making. Then, another algorithm takes automatic action by sending back signals representing anxiety and anger to the victim's brain. Subsequently, a series of verbal attacks via V2K, which were previously considered unnerving, begin.

This occurs until the EEG is obtained at the desired, pre-configured rhythm. In this case, the anger, fear and acute stress are the feelings that must constantly emerge, reflecting the up-to-date picture and backing the decisions to be taken by operators based on the reality of the new data. It happens in a loop until the target's state of stress reaches its peak. Consequently, the process of arguing — retorting, pouring out the pain of torture — against the operators using the internal (or silent) voice of the mind via SYNTELE is initiated in an organic and instinctive way. It can escalate to the point of screaming out loud. It then maintains an ideal behavior in which all the victim's forces are directed against the attacks composed of voices, which initiates an intense "squabble" among targets, operators and AIs. At this point, a full connection is established between the scheme and the target. If it persists, it'll completely exhaust the victim, undermining their mental and physical resistance, as well as leading them to a debilitating condition without precedent.

To aggravate the situation, it'll be worse if the target lives with someone — relatives, friends or spouses — who knows nothing about the attacks and participates in events of acute stress caused by the use of the AI, in which the target instinctively retorts the operators. That is, this sudden scene, in which the victim at first babbles and finally utters the same words in a shriek, is soon seen as a symptom of serious mental illness to those around them. This is another type of attack widely exploited. This conduct is foreseen and occurs as a result of only one sporadic counterattack among thousands that are yet to come while the target is under the control of this technology. It takes place through the pattern of EEG waveforms that apparently have only one objective. However, other wicked goals are hidden in the course of events, as the direct consequence of subsequent behavioral acts.

This abnormal behavior is one of the most expected by MKTECH operators. It maintains a high level of stress and it's also possible to glimpse the expected effects on the human brain (often unconsciously by the person) that is automatically prevented from getting involved with any other activity that is taking place in the environment around them. Furthermore, levels and intensity scales for certain emotions are gauged after capturing the default settings of the main feeling. The anger is an example of it. The individual can go from calm (zero) to displeasure, followed by an uncomfortable irritation to a violent tantrum, which is the ultimate goal. The top of the scale can be quickly reached using SYNTELE (Synthetic Electronic Telepathy).

The degree of tension is also verified. It can trigger a variety of evil impulses, generating a scalable measure of excitement, based on previous experiences that the MKTECH operators themselves have already created in the target's mind. Remember that the systems are at the center of what happens between the victim and operators, including the interaction between all these feelings and the perception of reality that always involves sensory input and internally stored memories of past experiences.

Some feelings, such as motivation, cannot be directly observed by algorithms. Motivation for us is characterized by a strong energy unleashed and directed towards achieving goals. The AI creates an estimated scenario based on many variables, including previous learning with external help from humans, indicating to the machine that the target is currently motivated, which helps the AI to understand this complex abstract feeling that is inherently human.

Neural biometrics and remote EEG work together, using each other's functions and routines in a cohesive module. Remote EEG is essential for the functioning of the entire MKTECH scheme. It's responsible for mapping the emotions and the current psychophysiological state of the target. Without it, the technology wouldn't work. It has an internal module that incorporates the so-called affective computing: a technology that allows computers to detect and react to human emotions. All these features are present in the Mind Control Technology.

The influence of this technology on the central nervous system is so great that it can change the mental state in seconds. It can also send signals directly to the Targeted Individual's mapped areas, which alters waves from the current normal state to another state in a crude way, projecting a previously recorded configuration, stimulating areas to synchronize their activity at these frequencies and generally leading the individual to constant fatigue, drowsiness, stress and sudden abnormal sensations that substantially change the psychophysiological state for no apparent reason. Moreover, the AI takes into account a deeper analysis of the current scenario, the so-called principle of contextualization. How the brain — as a whole

— responds to a sensory stimulus to produce a particular motor behavior depends on its global internal state at each moment. Thus, these continuous mappings of states are essential to MKTECH operators. They indicate to AIs what attitude to take in terms of psychotronic attack on the target at each moment of the day. They even point out the moment the victim is most susceptible to theft of information inside the mind, leading the brain to organize thoughts and access memories even with the target using mental countermeasures to divert access to certain memories, or by capturing negative states and intensifying them to a point where the target performs severe actions, such as mutilation, suicide or murder.

These are then the modules responsible for interacting with deeper, primary innate emotions: joy, sadness, fear, disgust, anger and surprise. And complex secondary ones, which are negative social emotions, such as envy, jealousy, guilt or shame, as well as background emotions (well-being, malaise, calm or tension).

Thanks to the algorithms and the EEG working around the clock, operators will have a constantly updated picture of all aspects of the target's life. The victim, in turn, has little chance against these people and artificial intelligences specialized in human ailments and emotional depths that attack their brain with a level of knowledge that should be strictly private.

Emotion modulates some integrated mechanisms, one of them is the intensity of the feeling. Level of tension, an emotional response generated by the situation that was awakened in the subject — the fight-or-flight response. The answer is given by the degree of excitement and based on previous experiences and interpretations — memory and thought. The system decodes this entire scenario and transforms it into visual data that helps humans to make decisions in order to best intensify the suffering of the Targeted Individual, leading them to an imminent collapse.

3.2.3 - Target location anywhere in the world. Neural GPS

One of the victim's greatest concerns is to understand how they're located around the world by terrestrial and space antennas that are part of the MKTECH scheme, as they'll be subject to attacks via

SYNTELE wherever they are. The telemetric EEG along with the Neural Remote Biometrics are responsible for analyzing the individual's activities and brain waves in real time. One of its sub-functions is to constantly send a signal — these settings — to remote computers via satellite for comparison with the neural biometric data pre-established in the database. Thus, they indicate to the system whether the target is really the target and their position on the planet using GPS triangulation. It's similar to your cell phone's GPS tracking system and geolocation. It's possible to locate a person wherever they are, even inside buildings, houses and common urban structures and some natural ones, such as caves and grottos. Afterwards, the position will make the satellite choose the area of the beam where the trigger will have priority, always keeping the quality of the connection.

3.3 - Remote Polygraph - The most efficient lie detector ever created

"Even if small behaviors resulting from meticulous observation reveal that someone is in fact lying, confirmation via REMOTE POLYGRAPH takes a new level in detecting lies. It's by far the most accurate 'Lie Detector' on the planet. It's practically impossible to keep any kind of secret, especially when you're taken aback and is unaware about the existence of this technology. Facing the remote polygraph unprepared is like seeing the truth 'flow through the air'."

— FelipeSSCA.

The act of lying — that is, asserting what is known to be false, or denying what is known to be true — is present in human life. For example, it's common to lie in informal conversations. Most commonplace lies serve to keep social relationships stabilized without generating stress. Thus, appearances are preserved by slightly distorting, exaggerating or mitigating the facts. Even nature does that: it likes to hide itself. Lying is part of our growth process like walking and talking. We learn how to lie between the ages of 2 and 5.

Obviously, I'm not referring to big lies, or lies that undermine others, that hide someone's true intentions to do harm, that hide serious crimes, or gain economic advantages. These lies are serious dishonesty and have social and legal relevance. In both cases, however, it's necessary to express thoughts in speech or in writing in order to know if the person is lying, since the truth is part of the thought, and the lie requires effort and mental flexibility to be maintained. So, the lie would have to be expressed in one of these acts. It's not possible to know whether the thought is a lie or not if it's confined to the brain alone. It can only be turned into a lie when expressed in behavior.

Well, it's not like that anymore. Thoughts trapped within your own mind that haven't been expressed are amenable to judgment to confirm whether or not they are true. The Miranda warning ("You have the right to remain silent") is basically over with the advent of listening to thoughts, which would be the act of talking to yourself internally, or the voice of the mind talking with other mental modules, such as the consciousness. Plato defines consciousness as the soul's dialogue with itself. In this dialogue — an involuntary and automatic process —, a set of psychic functions is generated and they manage our options in life and provide instructions about these choices, judgments of our decisions, actions and analysis of thought. This process is subject to being heard in the target's mind, making it impossible to hide any type of information.

During a police investigation, for example, authorities might be able to hear the silent conversation in the mind of the suspect even before he or she physically presents himself or herself to answer questions. By using this device, it's possible to conduct an interrogation based on data obtained illegally from the suspect's mind, raising questions that can easily elucidate the case. During the interrogation, it'll be clear whether or not the suspect was involved in the crime in question.

The EEG, Remote Polygraph, evaluates data coming from SYNTELE (Synthetic Electronic Telepathy), ERM (Electronic Remote Monitoring), V2K (Voice to Skull) and D2K (Dream to Skull). Together these modules form a virtually foolproof mechanism for detecting a lie. A lie that could be, for instance, the number of goals you scored in a match (just to impress a friend), distorted memories relived during a bar conversation or a heinous crime kept under lock and key inside a murderer's mind. It's unnecessary to open the key with this weapon; it simply gets to access the data before it's even locked away.

But, before we go any further, let's understand how the polygraph or "conventional lie detector" works. The polygraph is a device that measures and captures logs of various physiological variables while an interrogation is carried out. It's used to try to identify lies in a report. Basically, the device registers small physiological changes in situations given by the interrogator.

How normal polygraph tests work

* **Pretest** - it consists of an interview between the examiner and the examinee, when they get to know each other better. This may last about one hour. At this point, the examiner listens to the examinee's version of the events under investigation. While the examinee is answering the questions, the examiner also profiles the examinee. The examiner wants to see how the subject responds to the questions.

* **Design questions** - the examiner asks questions specific to the matter under investigation and reviews the questions with the examinee.

* **In-test** - the actual exam is conducted. The examiner asks 10 or 11 questions, only three or four of which are relevant to the matter or crime being investigated. The other questions are control questions. A control question is very general, such as "Have you ever stolen anything in your life?" — A kind of question that is so broad that almost no one can honestly answer with a 'no'. If the person answers 'no', the examiner can get an idea of the examinee's reaction when he or she is lying.

* **Post-test** - the examiner analyzes the data of physiological responses. If there are significant fluctuations in the results, it could mean that the subject has been deceptive, especially if the person displayed similar answers to a question that was asked repeatedly.

The human factor and the subjective nature of the test are two reasons why polygraph test results are rarely accepted in a court of law. Here are two ways an answer can be misinterpreted:

* **False positive** – the answer of a sincere person is determined deceptive.
* **False negative** – the answer of a deceptive person is reported as being truthful.

The remote lie detector is based on the recording of autonomous physiological reactions as the conventional one, but with the ability to observe the situation that provokes emotions in a deeper analysis of the direct reading of thoughts. Questions are carefully designed to receive an automatic response from the brain at any level. Once again, keep in mind that the Synthetic Electronic Telepathy is able to send questions to the mind without the possibility for the victim to ignore them and hear the response to the stimulus being processed, according to the content of the message. Critical questions can incite emotions in the target, if they have contact with the subject under discussion.

The respiratory rate, heart rate, blood pressure, skin temperature, sweating, increased blood pressure, gastric and intestinal changes, among other physiological effects, are captured. However, it's possible to "hear" the reason in your silent thought and real-time visual memories. That is, what caused certain autonomous behavior, what led you to the state of anxiety due to guilt or fear of being discovered, making the remote polygraph "almost" infallible. I use "almost" because we have to take into account the complexity and diversity of how each brain shapes itself during life. Furthermore, the way in which data is handled is unique. The smallest variation in the stimulus, or in the bioelectric mechanics, can completely alter the perception of this information. Your nature and your interpretation of certain events, when receiving this data, can pass through the subjectivity of observation and the universal human tendency to perceive only what is wanted and not what is really happening. So, each person has a different reaction to the Remote Polygraph. For much of the time, and in most surviving victims, however, the device is extremely efficient in a linear manner concerning the results, as presented here.

Another huge advantage over the conventional polygraph is the fact that it can extract data even from those who are devoid of emotions. The individual being questioned may have difficulty understanding the consequences of their act or perhaps their limbic systems don't operate properly. In this case, the ERM is the main source of brain information capture. One can still come across the data obtained in statements that are quite questionable. People, when describing their internal experiences, may or may not deliberately hide some information and misrepresent others. The remote polygraph easily resolves any doubts, as it clearly hears the thoughts regarding the question.

Signal capture in sensory receptors carry the most varied types of information to the central nervous system. As they travel, they cross synaptic stations, giving rise to conscious activities, a process that is beyond the person's control. Thus, it's impossible to prevent this message from reaching the operators in the same way that they reach the victim's "self" in the understanding of the information. Bear in mind that behavior is the product of the functioning of physiological mechanisms. Each corresponds to a basic organic structure — among them, of course, the thought.

In the case of victims turned into targets, who live 24 hours with their thoughts analyzed by intruders who measure various parameters, this process of verifying the occurrence of the truth happens every second with any thought — from the most banal to the most relevant. Operators use this constant data to fine-tune the perception and confirmation process with maximum accuracy, inferring reliable results as to whether or not the victim is lying in response to a question sent through the technology's transmissions. Lying usually requires a lot of brain resources. Crafting a lie and keeping it at every step is an arduous task for most people's minds. For a target who must maintain a lie 24 hours a day, including dream confirmations via D2K, it's just impractical.

The Remote Polygraph is based on a set of information collected by Telemetric EEG. Data analysis is able to categorically state whether a given thought is or isn't fiction, or some memory actually accessed, or whether the response to an act is half true, half imaginative creation or a copy of some event seen in a movie, for example. The target usually speaks internally (silent thought) about the origin of the memory, the name of the movie from which certain information was accessed. It's extremely difficult to "brake" the brain when it wants to access information quickly and leave a silent thought on the tip of the tongue, ready to be expressed in words. But we cannot blame our minds for being efficient, after all, that's their job. What we have to do is to prevent the use of this weapon indiscriminately so that we can once again have the right to think inwardly and freely.

Regarding the Remote Polygraph: the RP is able to use all current detection methods and distinguish active brain regions where memory is accessed. It checks if a given situation is the result of imagination or if the target actually went through it; if the matter is new or if, once processed, the memories relating to the facts raised will be accessed. If these memories are recognized, pre-stored memory areas will be recruited. That is, even if the individual creates a lie, the internal processes mapped cannot deny the truth. Implemented in any fragment of memory, in any situation — whether small or not — these tests are usually conducted to snatch something from the target that will embarrass them in the future, to be repeatedly revived in their memories, causing the greatest possible amount of emotional damage.

The foundation of the Remote Polygraph lies on the functional characteristics of the normal polygraph. However, the incredible ability to hear and see thoughts is added, thus culminating in an accurate analysis of every bit of thoughts that form an idea. In the MKTECH case, this is done remotely and without the victim's consent. The scheme steals from them a series of physiological sets intercepted by the main module — the Telemetric EEG.

The immediate organic reaction to a question coming from V2K is analyzed right after the sound transduction and the understanding of the message embedded in it by the cortical areas responsible for this process. In the last step, this response is "amplified" (reradiated) by the EMR (Electronic Mind Reading) at the speed of light. All these elements *upgrade* the Polygraph concept.

There's an old saying that says it's not possible to fool a polygraph, but rather the person who is trained and responsible for analyzing the data. In this case, the examiner has a lot of extra information that prevents the wrong conclusion in most cases analyzed.

Discomfort about a particular subject, including the neural visceral reactions that occur when the target encounters something unexpected, can be noticed. Since the victim has had every aspect of their life previously studied — their everyday thoughts monitored and recorded for months before the attack begins — the operators have an advantage. The means of defense are practically nil.

The initial massive data collection we've seen in previous chapters serve to calibrate the Remote Polygraph with the victim's profile; in other words, how they act in the face of the truth, and how they behave when lying on multiple levels. The neutral state is zero. It shouldn't evoke any kind of emotion and serves as a standard emotional reference point to compare with reactions followed by critical questions that are known to cause organic changes in the target. A white lie that cannot be captured by external signals — and even fat lies in which the target clearly has some physiological reaction detectable by the AI that works with the Polygraph — **serve as an input for creating a base of emotional signatures, cataloging positive or negative interference capable of generating changes in visceral functioning (the famous "butterflies in the stomach")**. At the deepest level, one looks at how the target interprets whether a given attitude is socially acceptable or not in their minds. Please note the degree of moral/psychic invasion that this weapon is capable of doing to people in a negative way.

As we know, targets cannot control their thoughts and are not used to having to restrain them that way — the operators make a fuss in their mind. They activate personal memories, situations, feelings and matters, which causes a great deal of trouble. It's as if the victim has several internal voices accessing their memories and measuring their reactions. It's similar to playing with a compromised operating system. This system, however, is your personality, your life, your "self", and it creates a ripple effect on everything and everyone around you. The victim needs years of training to be able to prevent operators from accessing memories, stealing information or embarrassing them with personal memories, opinions on a certain subject and urges inherent in every human being. This keeps the target constantly immersed in anxious thoughts.

As the internal conversation starts, natural thoughts that provoke small behavioral reactions — as an involuntary and uncontrollable reflex caused by the sympathetic nervous system —, such as blushing, shame or physical discomfort, will be a strong indication that the person knows about the subject raised by the operator, but using the remote polygraph. Besides, a number of elements also appear, including the vocalized thought and visual memories of the act itself.

This makes it impossible for anyone, even if relatively prepared, to deceive this modern polygraph.

The cowardice of this act refers to the unconscious way in which automatic processes in the brain concatenate the flow of thoughts, accessing memory, giving a judgment, propagating through executive faculties. This procedure isn't optional; it's used as the main source for obtaining information through the remote polygraph. Even after a lot of training, the brain still feels the urge to answer certain questions internally, playing tricks and revealing data about their private life due to the modern tactics of interrogation conducted inside the minds of anyone who becomes a target. One of these techniques is to use unconscious processes, a kind of brain automation to complete incomplete sentences and phrases.

For example: "Remember that day at so-and-so's house that you...". If the target was present at the person's house on that date, an unconscious response will complete the narrative. This tactic is used at all times for various types of stratagems: from confirming a certain detail about events to discovering more intimate data about the target and the respective event on a certain occasion, together with the analysis of the responses, giving opinions about a certain person, or simply to capture some low-level gossip to be used as SYNTELE attack ammo. This tactic is foolproof, as it's virtually impossible to stop the brain in its search for information. It's also possible to capture and use their opinions about the people close to them and embarrass the victim for thinking a certain way using the same method. Furthermore, one can find out whether the answer came from areas linked to creativity or areas linked to memory.

The sensation of having people "living" inside your head, filtering, diverting, monitoring and activating thoughts and memories is so unique that it doesn't even seem real.

Experiments captured in the face of the remote polygraph

One of the most valuable experiments conducted by operators, which takes place in a deeper, completely hidden layer of the scheme and using the background of torture, is precisely at what point in time and what mental resources the target used in order to no longer access information in memory. That is, how they finally manage to control their thoughts by not revealing more information to enemies. An arduous task, which is not 100% efficient, but leaves operators with a false positive or inconclusive result on whether a certain issue is true or false, or whether the victim participated in a certain situation or was somewhere else.

All torture victims are part of this complex experiment — as we will see in depth in the next volume. This is a new tactic to be employed: how people manage to hide information (their thoughts) inside their own mind, adding more parameters to the new modern defense protocols for certain key people in the industry, in intelligence and military agencies.

It's undeniable that the MKTECH as a whole — used within strict rules to solve crimes by the justice departments and to assist in interrogations — completely changes the criminal justice system, as we know it today. Take a suspect at a police station — asked to testify on a certain matter, a murder — for example. Once the suspect enters the interrogation room and the system is turned on in their mind, it'll be easy to find out if they participated in the crime and to what extent. By listening to the inner thoughts along with the telemetric EEG data, in a matter of minutes, or even before the person sits down, this information will already be known. However, what about the private thoughts that are generated throughout the process that have nothing to do with the crime itself? Who will analyze and completely violate the examinee's cognitive privacy? What if the person is innocent and another crime is detected in the process of reading thoughts? What if they remember an affair? What if they curse the investigators internally, using the voice of the mind? What if a thought concerning a secret job project suddenly came up? *What if...* What would the captured data look like?

Thoughts cannot be controlled as the act of speaking or writing can. These questions will be raised when this technology is regulated and laws governing thought hacking are created. I believe issues like this will soon be debated by society. Today, uncontrolled technology is capable of doing all of the above. Rest assured that it's practically inevitable to find out if the person is lying or not. After following standard procedures, investigators will likely have all the answers about the crime committed if they turn on the MKTECH system in the suspect's mind inside their home and capture the mass of data. This is a very complex subject that deserves a more accurate philosophical, social, technological, criminal and legal analysis. It can be covered in future books or articles.

What worries me today is the way it is used indiscriminately to satisfy the perverse desires of certain groups, which brings chaos and destruction to the lives of many Targeted Individuals. In addition, the situation may worsen in the future, becoming irreversible when it falls into the public domain. Nevertheless, there is hope. If the victim has already knowledge of the existence of the technology and techniques used by operators and torturers, the ability to defend themselves increases. This mitigates the primary negative consequences of dealing with the polygraph when used in a harmful way. The target will no longer be automatically stimulated by the voices that imitate their internal voice and trigger the process of accessing the thought desired by technology operators. But the victim may still reveal the truth in the face of constant psychotronic torture and through "tapped" channels: dreams, sounds, memories, thoughts.

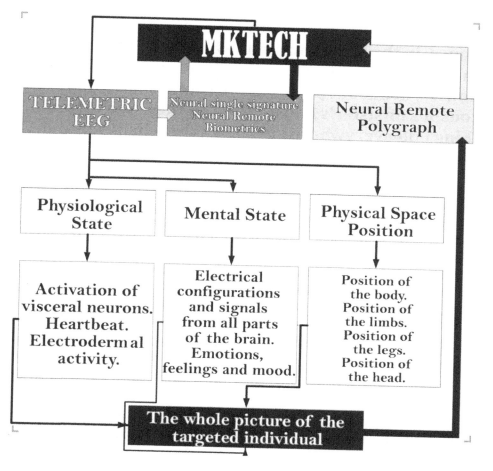

Figure 3.10 *Complete scheme of the systems and their tracking modules via EEG, remote polygraph support, mental and physiological states monitored and qualified.*

3.4 - How can they see me all the time, inside my house, in the bathroom, in the bedroom, at work, at the beach, in the park, on the street, at friends' houses, in distant places in the countryside, at my cousin's or sister's house? How do they "see" everything I do everywhere I go? They see what I see and they hear what I hear all the time

One of the issues that most causes outrage, perplexity and raises questions from targets is: how do operators get to know exactly everything about them in real time? They even know the position of each limb of their body in space and what they're currently doing behind closed doors. At this point you already know the answer. Data acquisition takes place through the various modules of this weapon, which include surveillance, defense and neuroelectronic attack. This set of tools creates an up-to-date picture through thorough monitoring. Based on this data, the system generates a reliable virtual image that is on a screen in a continuous manner and changes according to the person's real movements. At home, for example, one can tell if the target is sitting, standing or what activity they're engaging in with amazing accuracy.

A rough virtual image is formed. It's even possible to predict the movement of hands and fingers, of arms and legs, head and chest, and anticipate actions based on previous behaviors — responses to stimuli with the participation of the AIs who learned from interacting with the individual. Sophisticated equipment that captures a thermal signature — similar to a radar echo — can also be used to assist in this unprecedented invasion of privacy.

Let us not be surprised that these technologies are being gradually introduced into our society, without much uproar. Given the proper proportions with the military technology we've seen throughout this chapter, there's already something similar to that in the civilian world. MIT, for example, is developing a technology capable of seeing a person's position through a wall. The paper "Through-Wall Human Pose Estimation Using Radio Signals" calls RF-Pose "a solution that leverages radio signals to accurately track the 2D human pose through walls and obstructions." They use radio signals to identify — with 83% accuracy — someone in a line of 100 people. Initiatives like these are beginning to emerge in many sectors, consistently supported by radio waves and AI technology. This type of technology is already developed, and is similar to the one used for positional mapping of the target in their home through radio wave interactions.

The technology

EMR - Electronic Mind Reading. One of the main systems responsible for "wiretapping" all cognitive functions, subdivided into:

EMRi - Electronic Mind Reading (images): responsible for amplifying visual thoughts (of images), capturing and sending them to a computer program, demodulating and reproducing the target's mental images.

EMRv - Electronic Mind Reading (vocalized): responsible for amplifying internal/vocalized thoughts, capturing and sending them to a computer program, demodulating and reproducing the target's thought content.

The components **EMRo** - Electronic Mind Reading (optical) and **EMRa** - Electronic Mind Reading (auditory) contribute to providing an environmental context, in addition to calibrating the abuse and invasion of privacy by listening to everything that goes on in the victim's' auditory system and the images that are received straight from their vision. That's why operators see what the target sees and listen to what the target hears. The famous "*I see what you see; hear what you hear*". Disturbing, to say the least.

All of these devices operate in perfect sync, at full throttle. They tell us why the target is tracked all the time — every second of their life is shared immediately and in detail. Thus, the Synthetic Electronic Telepathy (**SYNTELE**), which devastates thoughts by capturing any process that takes place in the mind, also joins the harassment and stalking team. And there is the Intracranial Voice (**V2K**) that torments the target within their own intimacy and drive them crazy by accessing intentions and inducing thoughts that give clues to what activity the target is currently engaging in.

We cannot forget the Synthetic Electronic Dream (**D2K**) that creates a world, an entire alternate reality, whenever the target goes to sleep. Each night a different hideous nightmare reveals a lot about the victim and forever alters visual memories. As we learn about all the technologies that make up MKTECH, it's evident how operators know every aspect of a target's life. They see and hear everything that the victim does.

Epilogue

We've now come to the end of the first volume of the book. Here I've detailed and presented modules that make up the scheme that forms a weapon powerful enough to reach and end the life of everyone on the planet. We've got a little taste of how it works. We've learned about the division of each module responsible for a specific brain area, which together form the ultimate electromagnetic neural weapon. This weapon works in an unprecedented way in the human mind, invading and penetrating the deep layer of interaction of cortical modules. It interferes and directly alters processes that were previously restricted to our inner side and the result of their natural relationship with the environment, artificially shaping our experiences and the interpretation of reality as it's presented to us.

As we've learned a little more about MKTECH, it became clear that the only place we think was safe to keep our secrets no longer exists. Our mind is easily hackable and has no defenses against it. This is a consummate knowledge that will force society to undergo unique changes in the event of the popularization of the weapon in the future. Everything we've learned was just the kickoff for a full understanding of this dark and unbelievable universe of an extremely powerful technology.

All of this takes us to the next volume of the book (Volume 2) in which even more thorny issues will be discussed. You'll learn about advanced techniques for the use of these neural weapons, in what context they're used, the history behind their emergence and the illegal experiments conducted on humans by governments with the codename MK-ULTRA that gave rise to these weapons and extend to the present day.

As you read the chapters, you'll understand how persecution and torture based on real cases work, the advanced tactics employed in the use of psychotronic weapons that involve the thoughts of the targets and the theft of information and brainwashing techniques. You'll also understand how children and animals are vulnerable to these weapons and how a child's growth and development is affected in an attack of this magnitude. If you have a child, this book is a must-read!

You'll also learn about undetectable tactics that are used to defraud civil service competitive examinations, the worldwide terrorist attacks using these weapons, electromagnetic warfare, artificial intelligence at the service of MKTECH, satellites prepared to hack the human mind and much more.

www.invasionandmindcontrol.com

Glossary I

Glossary I is not in alphabetical order. In fact, it's arranged in a way that helps the visualization of major neuroelectronic technologies as a whole.

MKTECH - Mind Control Technology: complete system covering all technologies or modules below.

EMR - Electronic Mind Reading.

EMRi - Electronic Mind Reading (images): subsystem which is part of a complex scheme that uses a series of electronic devices to capture, amplify and decode the content of the electrical signals from neural networks responsible for mental images or visual memory of thoughts.

EMRv - Electronic Mind Reading (vocalized): subsystem which is part of a complex scheme that uses a series of electronic devices to capture, amplify and decode the content of the electrical signals from neural networks responsible for the vocalization of thoughts (the voice of the mind).

EMRa - Electronic Mind Reading (auditory): subsystem which is part of a complex scheme that uses a series of electronic devices to capture, amplify and decode the content of the electrical signals from neural networks responsible for hearing. One can hear everything the target is listening.

EMRo - Electronic Mind Reading (optical): subsystem which is part of a complex scheme that uses a series of electronic devices to capture, amplify and decode the content of the electrical signals from neural networks responsible for the sight. It's possible to see images that the target visually processes.

V2K - "Voice to Skull", Intracranial Voice, Microwave Voice or Microwave Hearing Effect: system capable of inserting voices and sounds directly into the target's mind. These are microwave transmissions that are demodulated by the target's brain and are indistinguishable from the sound picked up by the ear. There are no outside noises; people around them cannot hear the sounds, only the target. It's a modern version of the V2K (v2.0) that has all the qualities of its predecessor, but with improvements in sharpness and the capacity of "blending" with the noises that come naturally through auditory afferent pathways. The target hears voices "inside" any noise that is naturally picked up. They hear voices within the noise of engines, appliances, airplane turbines, etc. The noise energy "amplifies" the microwave voices in the target's brain. These voices now have the same decibels of the sound that comes at all times. Another added quality is that the microwave sound is interpreted by the target according to the environment. That is, microwave voices "receive" the acoustic signature of the environment in which the target is immersed. If the target is in a garage, every microwave voice will "receive" the natural echo of the place or the identical reverberation of the environment. This effect is largely responsible for creating mental confusion. It makes the target think that people are chasing them wherever they go.

SYNTELE - Synthetic Electronic Telepathy: consists in the use of the EMRvi – Electronic Mind Reading (vocalized/images) to extract information — content of thoughts — and the V2K to insert data — voices and sounds — in order to activate neural processes connected to hearing. Then, a totally silent conversation between the technology operators and the targeted individual takes place. In other words, it's possible to send messages to the operator with the thought alone and receive a response via V2K without anyone in the vicinity being able to hear what is being communicated to the mind, nor the content of the thought being amplified and transmitted. It closes, therefore, a complete communication cycle. It's also responsible for inserting /sending images to the targets' visual thoughts that are similar to what we visualize in our minds when we think of a material object.

D2K - Synthetic Electronic Dream or SLEEPING BEAUTY: is a MKTECH module with the very specific purpose of completely mastering cognitive processes while the target is unconscious (sleeping). Operators practically have complete control over the graphic content that will be displayed within the individual's mind. The images are sent through electromagnetic transmissions to be demodulated by the brain of the mapped individual the moment they leave the waking state and begin to sleep. The

Targeted Individual dreams about the content transmitted by operators. These images are captured by the brain at any stage of sleep, but the greatest interaction takes place during REM sleep.

RNM - Remote Neural Monitoring: this module is of great importance for conducting the entire MKTECH scheme, due to several unique characteristics composed of three distinct elements converging towards a purpose that is achieved in a joint effort: to monitor all possible raw physiological data as well as all electrical brain wave patterns, which indicate the individual's current physical and mental state, assisting the torture and physiological (reactive) responses that are of great value to the tests conducted by the operators.

EEG-Based Biometrics: feedback signal sent to the target's brain, or re-radiation of electromagnetic energy transmitted to the brain. It contains all the unique electrical characteristics of each sample. Several mathematical models and algorithms were created by scientists involved in the evolution of this weapon to individualize the brain waves mapped more and more precisely. The BCI is responsible for receiving this information in real time, quantifying it appropriately and forwarding it to another responsible algorithm or subsystem, translating it into a unique pattern for each individual on the planet.

Telemetric EEG: Electronic Brain Link (EBL) or Remote Neural Monitoring. It's able to accurately gauge various complex psychophysiological states of the targets, measuring, stimulating and distorting feelings and emotions, such as fear. It calculates the approximate data, such as heart rate, changes in neurons in the bowel ("butterflies in the stomach") and blushing. It's possible to create a virtual avatar that displays the body position in space in real time of all target's limbs in a graphical representation with the electrical feedback data obtained. Thus, they visualize in which position the body is in a defined period of time. They remotely monitor brain waves and electrical changes, biomagnetic analysis system similar to ECG, EEG, EOG, EMG, EGG, Respiration Rate, Pulse Rate, Temperature, Impedance Cardiography and Electrodermal Activity. These data can be captured or inferred with a satisfactory margin of approximation if compared to the equivalent of using electrodes, as well as brain functions and firing patterns of neurons involved in certain behaviors over a period of time.

Neural GPS: The telemetric EEG along with the Neural Remote Biometrics are responsible for analyzing the individual's activities and brain waves in real time. One of its sub-functions is to constantly send a signal to remote computers via satellite for comparison with the neural biometric data pre-established in the database. Thus, they indicate to the system whether the target is really the target and their position on the planet using GPS triangulation — similar to your cell phone's GPS tracking system and geolocation. It's possible to locate a person wherever they are, even inside buildings, houses and common urban structures and some natural ones, such as caves and grottos. Afterwards, the position will make the satellite choose the area of the beam where the trigger will have priority, always keeping the quality of the connection.

Remote Polygraph: it evaluates data coming from EEG, SYNTELE, ERM, V2K and D2K. Together these modules form a virtually foolproof mechanism for detecting a lie. A lie that could be, for instance, the number of goals you scored in a match (just to impress a friend), distorted memories relived during a bar conversation or a heinous crime kept under lock and key inside a murderer's mind. It's unnecessary to open the key with this weapon; it simply gets to access the data before it's even locked away. It analyzes the thoughts that are expressed.

Glossary II

In alphabetic order:

AI – Artificial Intelligence.
BCI – Brain-Computer Interface.
BMI – Brain-Machine Interface.
Brain Net/Web – internet or network where data extracted from the mind (thoughts) travel.
Dark Mind Web – dark internet of thoughts.
DARPA – Defense Advanced Research Projects Agency.
Deep Brain Web – deep internet of thoughts.
DEWs – directed-energy weapons.
DOD – The United States Department of Defense.
EEG – Electroencephalography.
Electronic Psy Ops – electronic psychological warfare operations.
Electronic Psychological Warfare - unconventional electronic and psychophysiological war tactics using psychotronic electromagnetic weapons.
Gang-Stalking or Organized Gang-Stalking – they're nothing more than the human representation of the Mind Control Technology.
KUBARK – codename used by the CIA to refer to itself.
LGN – Hacking: hacking focused on breaching the Lateral Geniculate Nucleus electrical data.
MGN – Hacking: hacking focused on breaching the Medial Geniculate Nucleus electrical data.
MK-ULTRA – 1950's mind control and research program.
MKULTRA 2.0 – modern research and mind control program with psychotronic or neuroelectronic weapons.
Neural claustrophobia – suppressed thoughts, fear of thinking, need to drive out invaders, thinking without being heard by invaders.
Neural Phone – classified intelligence apparatus, precursor of SYNTELE.
Neural Satellites – neuroelectronic attack satellites.
Neuroelectronic Warfare – war in which enemies use electronic means to invade and destroy the human mind.
Neuromatrix – Hacking: hacking focused on breaching the Neuromatrix electrical data.
NLW – non-lethal weapons.
OPS – Organized Professional Stalkers or POS – Professional Organized Stalking together with OPT are a group or groups of people responsible for the content of torture that will be imposed on the victim in order to carry out the "stalking" and still maintain constant surveillance.
OPT – Organized Professional Torturers.
POWs – prisoners of war.
Psychotronic/Electromagnetic/Neuroelectronic Weapons – electromagnetic weapons designed to interact with, manipulate and destroy the human brain.
Psychotronics – neuroelectronic weapons.
Radar – Radio Detection and Ranging.
REM – Rapid Eye Movement.
SATAN – Silent Assassination Through Adaptive Networks.
Targeted Individual – victim attacked with neuroelectronic/electromagnetic weapons.
TI – Targeted Individual
Torture for fun – illegal surveillance, electronic harassment, sexual harassment.
V2K - Voice to Skull (Voice of God) Weapons.
Winter Soldier – modern version of the remote killer.

About the Author

Felipe Saboya de Santa Cruz Abreu

Felipe Saboya de Santa Cruz Abreu was born in Rio de Janeiro and pursues a career in computer science as a systems analyst/developer. He graduated in the Undergraduate Information Systems Program and completed the Post-Graduate Program in Computer Forensics & Cyber Expertise. Since 2012, the author has been engaged in the study of psychotronic/neuroelectronic technologies and their effects on the human mind.

www.invasionandmindcontrol.com

References

"I hunt men, as a Corsair does, but not in order to sell them into slavery, but to carry them off with me into liberty."
— **Friedrich Nietzsche.**

A 200–2700 MHz 2-Arm Conical Spiral Antenna. (n.d.). Superkuh. Retrieved August 26, 2021, from http://superkuh.com/conical-spiral-antenna.html

A Física do Rádio. (n.d.). A Física do Rádio [The Physics of Radio]. http://fisica3ufrb.blogspot.com.br

A Origem do Radar: A Origem das Coisas [The Origin of Radar: The Origin of Things]. (n.d.). A Origem das Coisas. Retrieved August 24, 2021, from http://origemdascoisas.com/a-origem-do-radar/

AbcMed. (2016, November 14). *Fases da infância - como elas são? Quais as mudanças envolvidas?* [Child development stages - what are they like? What changes are involved?]. https://www.abc.med.br/p/saude-da-crianca/1280663/fases+da+infancia+como+elas+sao+quais+as+mudancas+envolvidas.htm

About RTL-SDR. (n.d.). RTL-SDR. Retrieved August 25, 2021, from https://www.rtl-sdr.com/about-rtl-sdr/

Acústica [Acoustics]. (2021, April 8). In *Wikipedia*. https://pt.wikipedia.org/wiki/Ac%C3%BAstica

Agência Nacional de Informação Geoespacial [National Geospatial-Intelligence Agency]. (n.d.). In *Wikiwand*. https://www.wikiwand.com/pt/Ag%C3%AAncia_Nacional_de_Informa%C3%A7%C3%A3o_Geoespacial

All Identified Signals. (2020, April 2). SIGIDWIKI - Signal Identification Guide. Retrieved August 26, 2021, from https://www.sigidwiki.com/wiki/Database

ANATEL. (n.d.). *Atribuição de Faixas de Frequências no Brasil* [Allocation of Frequency Bands in Brazil]. Retrieved August 25, 2021, from https://www.anatel.gov.br/Portal/verificaDocumentos/documentoVersionado.asp?numeroPublicacao=&documentoPath=radiofrequencia/qaff.pdf&Pub=&URL=/Portal/verificaDocumentos/documento.asp

Andreas Spiess. (2019, September 8). *#286 How does Software Defined Radio (SDR) work under the Hood? SDR Tutorial* [Video]. YouTube. https://www.youtube.com/watch?v=xQVm-YTKR9s&list=PLus_DAOVXauZp4O2szPwzl2S0Ra1Gjypk&index=2&t=0s

Ano Internacional da Geofísica [International Geophysical Year]. (n.d.). *In Wikiwand.* https://www.wikiwand.com/pt/Ano_Internacional_da_Geof%C3%ADsica

Aprendizado de máquina [Machine learning]. (n.d.). *In Wikiwand.* https://www.wikiwand.com/pt/Aprendizado_de_m%C3%A1quina

Aranha, M., & Martins, M. (2003). *Filosofando: introdução à filosofia* [Philosophizing: introduction to philosophy] São Paulo, SP: Moderna.

Arnal, L., Kleinschmidt, A., Spinelli, L., Giraud, A., & Mégevand, P. (2019, August 14). *The rough sound of salience enhances aversion through neural synchronisation.* Nature Communications. https://www.nature.com/articles/s41467-019-11626-7?error=cookies_not_supported&code=7833324a-a232-4bf3-9358-11f6569e2169

Associated Press. (2017, February 17). *Esteban Santiago declared to be mentally competent.* Mail Online. http://www.dailymail.co.uk/news/article-4230110/Airport-shooting-suspect-Florida-federal-court.html

Association for Diplomatic Studies & Training. (2013, September 17). *Microwaving Embassy Moscow — Another Perspective*. ADST. https://adst.org/2013/09/microwaving-embassy-moscow-another-perspective/

Atores do Controle Remoto da Mente [Remote Mind Control Actors]. (n.d.). Google Sites. Retrieved August 25, 2021, from https://sites.google.com/site/controlemental/home/atores-do-controle-fisico-da-mente

Backdoor. (2020, February 17). *In Wikipedia.* https://pt.wikipedia.org/wiki/Backdoor

BBC News Brasil. (2018, January 29). *O distúrbio que leva uma mulher a conviver com cinco vozes em sua cabeça* [The disorder that causes a woman to live with five voices in her head]. https://www.bbc.com/portuguese/geral-42827481

Bear, M., & Paradiso, M. (2008). *Neurociências: desvendando o sistema nervoso* [Neurosciences: unraveling the nervous system]. Porto Alegre, RS: Artmed.

BEC CREW. (2016, June 23). *Scientists Have Invented a Mind-Reading Machine That Visualises Your Thoughts*. ScienceAlert. https://www.sciencealert.com/scientists-have-invented-a-mind-reading-machine-that-can-visualise-your-thoughts-kind-of

Bechara, E. (2015). *Módulo Sistema Nervoso – Neuroanatomia Funcional* [Nervous System Module – Functional Neuroanatomy]. Moderna Gramática Portuguesa. Rio de Janeiro, RJ: Nova Fronteira.

Benson, T. (n.d.). *With "BrainNet," scientists develop tech for brains to communicate directly*. Inverse. Retrieved August 26, 2021, from https://www.inverse.com/article/60596-brain-internet-connect-thoughts-messages?utm_campaign=inverse&utm_content=1572623892&utm_medium=owned&utm_source=facebook&fbclid=IwAR2EpyprCeA6Ki887Am14aYKa-Jyll7xlPtmF9towIQx9ehRbFJkwFWaYPIQ&refresh=49

Bergamini, C., & Tassinari, Rafael. (2008). *Psicopatologia do comportamento organizacional: organizações desorganizadas, mas produtivas* [Psychopathology of organizational behavior: disorganized but productive organizations]. São Paulo, SP: Cengage Learning.

Bhattacharjee, Y. (2018, January). *A ciência do bem e do mal* [The science of good and evil]. *National Geographic Brasil*, 214.

Big data. (2021, August 14). *In Wikipedia.* https://pt.wikipedia.org/wiki/Big_data

Bisi, G., Braghirolli, E., Nicoletto, U., & Rizzon, L. (2015). *Psicologia Geral* [General Psychology]. Petrópolis, RJ: Vozes.

Boric-Lubecke, O., Lubecke, V., Droitcour, A., Park, B.-K., & Singh, A. (2016). *Doppler Radar Physiological Sensing.* Hoboken, New Jersey: Wiley.

Braga, N. (2015, June 14). *Como funciona o Radar (ART154)* [How radar works (ART154)]. Instituto Newton C. Braga. https://www.newtoncbraga.com.br/index.php/como-funciona/10739-como-funciona-o-radar-art154

Braga, N. (n.d.). [Instituto Newton C. Braga]. Instituto NCB. https://www.newtoncbraga.com.br/index.php/eletronica/52-artigos-diversos/4261-art587

Braz Júnior, D. (2015, October 25). *Luz: onda ou partícula?* [Light: wave or particle?]. Tilt UOL. https://fisicanaveia.blogosfera.uol.com.br/2015/10/25/luz-onda-ou-particula/

Brown, G. (2001). *Radio and Electronics Cookbook.* Oxford, UK: Newnes. https://doi.org/10.1016/C2009-0-25079-3

Buzzi, A. (2001). *Filosofia para principiantes: a existência humana no mundo* [Philosophy for beginners: human existence in the world]. Petrópolis, RJ: Vozes.

Canídeos [Canidae]. (2021, May 2). *In Wikipedia.* https://pt.wikipedia.org/wiki/Can%C3%ADdeos

Carvalho, L. (n.d.). *Mundo da Rádio – Quando a potência não é tudo* [Mundo da Rádio – When power isn't everything]. Mundo da Rádio - O universo da rádio, na Internet. Retrieved August 24, 2021, from http://www.mundodaradio.com/artigos/quando_potencia_nao_e_tudo.html

CatsVsDogs. (2010, October 28). *EmoLens – Control Flickr with your Thoughts with the Emotiv EPOC headset* [Video]. YouTube. http://youtu.be/E9_XZlHoSp0

Catterall, W. (2011, August 3). *Voltage-Gated Calcium Channels.* PubMed Central (PMC). https://www.ncbi.nlm.nih.gov/pmc/articles/PMC3140680/

CBS News. (2017, January 9). *Esteban Santiago, Fort Lauderdale airport shooting suspect, makes first court appearance, is denied bond.* https://www.cbsnews.com/news/esteban-santiago-fort-lauderdale-airport-shooting-first-court-appearance-denied-bond/

CBS. (2019, September 1). *Brain trauma suffered by U.S. diplomats abroad could be work of hostile foreign government.* [Video]. https://www.cbs.com/shows/60_minutes/video/J53FM8S0J_qLalL8_lYk_nRcZHNkajaD/brain-trauma-suffered-by-u-s-diplomats-abroad-could-be-work-of-hostile-foreign-government/

Cérebro & Mente. (2003). Cérebro & Mente [Brain & Mind]. https://cerebromente.org.br/

Chaves, M. (1993). *Memória humana: aspectos clínicos e modulação por estados afetivos* [Human memory: clinical aspects and modulation by affective states]. Periódicos Eletrônicos em Psicologia. Faculdade de Medicina – UFRGS. http://pepsic.bvsalud.org/scielo.php?script=sci_arttext&pid=S1678-51771993000100007

Cheat [Cheating in video games]. (2021, April 26). In *Wikipedia.* https://pt.wikipedia.org/wiki/Cheat

CIA cryptonym. (2021, July 22). *In Wikipedia.* https://en.wikipedia.org/wiki/CIA_cryptonym

CIA cryptonym. (n.d.). *In Wikiwand.* https://www.wikiwand.com/en/CIA_cryptonym

Ciência Todo Dia. (2018, June 1). *Por Que Precisamos da Dualidade Onda-Partícula?* [Video]. YouTube. https://www.youtube.com/watch?v=CgY_zBuK2Cw

Ciência Todo Dia. (n.d.). *Home* [Ciência Todo Dia]. YouTube. Retrieved August 24, 2021, from URL https://www.youtube.com/channel/UCn9Erjy00mpnWeLnRqhsA1g

CINDACTA. (n.d.). DECEA – Departamento de Controle do Espaço Aéreo [DECEA - Department of Airspace Control]. https://www.decea.gov.br/?i=unidades&p=cindacta-i

ClintMclean74 / SDRSpectrumAnalyzer. (n.d.). GitHub. Retrieved August 25, 2021, from https://github.com/ClintMclean74/SDRSpectrumAnalyzer

CNN. (2018, September 2). *Microwaves suspected in attacks on US diplomats in Cuba and China* [Video]. YouTube. https://www.youtube.com/watch?v=Su0uc5UvLvg&feature=youtu.be&fbclid=IwAR2tEUb9WV6c9RxWkzinKW3BTWJ4qqHKJ_Lm3mopa-lrGgs9MraexvLyfwM

Costa, E. (2003). *Acústica técnica* [Technical Acoustics]. São Paulo, SP: Blucher.

Costa, E. (2009). *Eletromagnetismo: teoria, exercícios resolvidos e experimentos práticos* [Electromagnetism: theory, solved exercises and practical experiments]. Rio de Janeiro, RJ: Ciência Moderna.

Cristino, G. (n.d.). *Estrutura e Função do Córtex Cerebral* [Structure and Function of the Cerebral Cortex]. Gerardo Cristino. http://gerardocristino.com.br/novo-site/aulas/neurologia-neurogirurgia/cortexcerebral.pdf

Gattass, R. (2000, January 15). *O Pensamento - Mapeamento de Imagens por Ressonância Magnética Nuclear Funcional* [Thoughts: Image Mapping by Functional Nuclear Magnetic Resonance]. Cérebro & Mente. https://www.cerebro-mente.org.br/n10/mente/pensamento1.htm

Curso em Vídeo. (2016, February 19). *Curso Word #01 - Apresentação do Curso de Word 2016* [Video]. YouTube. https://www.youtube.com/watch?v=CgFzmE2fGXA

Curso em Vídeo. (n.d.). *Home* [Curso em Vídeo]. YouTube. Retrieved August 25, 2021, from URL https://www.youtube.com/channel/UCrWvhVmt0Qac3HgsjQK62FQ

Davidoff, L. (2010). *Introdução à psicologia* [Introduction to Psychology]. São Paulo, SP: Pearson.

de Carvalho, O. (2013). *Aristóteles em Nova Perspectiva - Introdução a Teoria dos Quatro Discursos* [Aristotle in New Perspective - Introduction to the Theory of the Four Discourses]. Campinas, SP: Vide Editorial.

de Godói, A. (2010). *Detecção de potenciais evocados P300 para ativação de uma interface cérebro-máquina* [Brain-computer interface based on P300 event-related potential detection]. Master's thesis, Universidade de São Paulo. Biblioteca Digital USP. https://doi.org/10.11606/D.3.2010.tde-19112010-115232

DEFCONConference. (2015, December 8). *DEF CON 23 - BioHacking Village - Alejandro Hernández - Brain Waves Surfing - (In)security in EEG* [Video]. YouTube. https://www.youtube.com/watch?v=c7FMVb_5SBM&list=PLus_DAOVXauYgyIgn-I6_7u2gUQEb-Rgu&index=42&t=0s

Delgado, J. (1969). Radio Stimulation of the Brain in Primates and Man Fourth Becton, Dickinson and Company Oscar Schwidetzky Memorial Lecture. *Anesthesia & Analgesia*.

Dell'isola, A. (2018). *Mentes Geniais* [Genius Minds]. Universo dos Livros.

Diario do Centro do Mundo. (2013, September 17). *"Bom rapaz" e "impulsivo: quem é Aaron Alexis, o atirador da base naval de Washington* ['Impulsive and good-natured': who is Aaron Alexis, the Washington Navy Yard shooter]. DCM. https://www.diariodocentrodomundo.com.br/bom-rapaz-e-impulsivo-quem-aaron-alexis-o-atirador-da-base-naval-de-washington/

Dormehl, L. (2018, February 26). *This A.I. literally reads your mind to re-create images of the faces you see*. Digital Trends. https://www.digitaltrends.com/cool-tech/university-of-toronto-mind-reading-ai/?fbclid=IwAR2YHymeiPRewKtVPfLkgxjPA4jkKdPLBdlGY0PWL8PhSwH9ZknHwvP4hP0

Dunning, B. (2018, February 27). *The Boy Who Thought He Was Reincarnated*. Skeptoid. https://skeptoid.com/episodes/4612

Edminister, J. (2013). *Eletromagnetismo – Col. Schaum* [Electromagnetism – Schaum Collection]. Porto Alegre, RS: Bookman.

Electronic torture, Brain zapping, Cooked alive, Electromagnetic mind control, Electromagnetic murder, Electromagnetic torture, Electronic murder, Microwave murder, Microwave torture, Organized murder, No touch torture, People zapper | STOPEG. (2008). Electronic Torture. https://www.electronictorture.com/

Elwood, J. (2012, November 14). *Microwaves in the cold war: the Moscow embassy study and its interpretation. Review of a retrospective cohort study*. Environmental Health. https://ehjournal.biomedcentral.com/articles/10.1186/1476-069X-11-85

EMOTIV | Brain Data Measuring Hardware and Software Solutions. (n.d.). EMOTIV. Retrieved August 25, 2021, from https://www.emotiv.com/

Espectro eletromagnético [Electromagnetic spectrum]. (n.d.). *In Wikiwand.* https://www.wikiwand.com/pt/Espectro_eletromagn%C3%A9tico

Esquadrão do Conhecimento. (n.d.). *Como funcionam os telefones celulares?* [How do cell phones work?]. Retrieved August 26, 2021, from https://esquadraodoconhecimento.wordpress.com/ciencias-da-natureza/fisica/como-funcionam-os-telefones-celulares/

Exame. (2017, November 2). *O poder do conhecimento* [The power of knowledge]. *Exame*, 1149.

Farquhar, G. (n.d.). *Protection from Neuro-Electromagnetic Frequency Mind Control Weapons*. Mark Jacobs. Retrieved August 25, 2021, from https://www.jacobsm.com/projfree/protection.html

Física com Douglas Gomes. (2019a, August 4). *Coaching quântico não tem embasamento científico. Você sabe o que é quântica?* [Video]. YouTube. https://www.youtube.com/watch?v=oRXNbnjD85U

Física com Douglas Gomes. (2019b, August 19). *Ondas e telecomunicações numa visão para ENEM (20 h) | Física com Douglas* [Video]. YouTube. https://www.youtube.com/watch?v=O5YqSWBKo2E&list=PLus_DAOVXauYgyIgn-I6_7u2gUQEb-Rgu&index=17

Fisica Universitária. (2016, September 23). *Eletromagnetismo – Espectro Eletromagnético* [Video]. YouTube. https://www.youtube.com/watch?v=-C2erXakQlQ

Fonética [Phonetics]. (n.d.). Portal Educação. Retrieved August 24, 2021, from https://www.portaleducacao.com.br/conteudo/artigos/educacao/fonetica/23454

Fort Lauderdale airport shooting. (n.d.). *In Wikiwand.* https://www.wikiwand.com/en/2017_Fort_Lauderdale_airport_shooting

Freedom of Information Act Electronic Reading Room. (n.d.). Freedom of Information Act. Retrieved August 24, 2021, from https://www.cia.gov/readingroom/

Frequência extremamente baixa [Extremely low frequency]. (2019, March 6). *In Wikipedia.* https://pt.wikipedia.org/wiki/Frequ%C3%AAncia_extremamente_baixa

Frequency-hopping spread spectrum. (n.d.). *In Wikiwand.* https://www.wikiwand.com/en/Frequency-hopping_spread_spectrum

Frey, A. (1993, February 1). *Electromagnetic field interactions with biological systems.* Federation of American Societies for Experimental Biology. https://faseb.onlinelibrary.wiley.com/doi/epdf/10.1096/fasebj.7.2.8440406

Fuentes, D., Malloy-Diniz, L., de Camargo, C., & Cosenza, R. (2014). *Neuropsicologia: teoria e prática* [Neuropsychology: theory and practice]. Porto Alegre, RS: Artmed.

Gallagher, J. (2019, April 11). *'De olhos fechados não visualizo nada': criador de método que revolucionou animação gráfica 3D não consegue formar imagens mentalmente* ['With my eyes closed, I cannot see anything': creator of a revolutionary 3D animation method, he cannot form images mentally]. BBC News Brasil. https://www.bbc.com/portuguese/geral-47866082?ocid=socialflow_facebook&fbclid=IwAR2vlMgNlVcRrJb9-slhDvWe97Bwya8GhWIHQObJkHDQiEFPdtzzny4CpXM

Golomb, B. (n.d.). *Diplomats' Mystery Illness and Pulsed Radiofrequency/Microwave Radiation.* Square Space. https://static1.squarespace.com/static/58fa27103e00bed09c8eac2c/t/5b7f95930e2e7262c9be0455/1535088022263/Cuba+2018-08-23c+-NEJM.pdf

Gomes, A. (2013). *Telecomunicações: Transmissão e Recepção AM-FM - Sistemas Pulsados* [Telecommunications: AM/FM Transmission and Reception - Pulsed Systems]. São Paulo, SP: Érica.

Graham-Rowe, D. (2002, May 1). *"Robo-rat" controlled by brain electrodes*. New Scientist. https://www.newscientist.com/article/dn2237-robo-rat-controlled-by-brain-electrodes/

Grinberg, E. (2018, April 24). *Travis Reinking: What we know about the Waffle House shooting suspect*. CNN. https://edition.cnn.com/2018/04/22/us/travis-reinking-waffle-house-shooting/index.html?fbclid=IwAR0a6C2ec8DThMOm3kCvuvSPlef_997URRRMMqCVplAHS5frFiWsvPaRFso

Grossman, N., Bono, D., Dedic, N., Kodandaramaiah, S., Rudenko, A., Suk, H.-J., Cassara, A., Neufeld, E., Kuster, N., Tsai, L.-H., Pascual-Leone, A., & Boyden, E. (2017, June 1). *Noninvasive Deep Brain Stimulation via Temporally Interfering Electric Fields*. Cell Press journal. https://www.cell.com/cell/fulltext/S0092-8674(17)30584-6?fbclid=IwAR0wXhavBMUFOhOB-TLjufXAg1Zaed4lWNBaAyTla6u62_1lkgUrgWIpgW0Q

Guia de CFTV. (2020, June 29). *Qual a Diferença Entre CCD e CMOS* [What's the difference between CCD and CMOS?]. https://www.guiadecftv.com.br/qual-diferenca-entre-ccd-e-cmos/

HackRF. (n.d.). Great Scott Gadgets. Retrieved August 25, 2021, from https://greatscottgadgets.com/hackrf/

Hambling, D. (2008, July 6). *The Microwave Scream Inside Your Skull*. Wired. https://www.wired.com/2008/07/the-microwave-s/

Haykin, S. (2017). *Redes Neurais: Princípios e Prática* [Neural Networks: Principles and Practice]. Porto Alegre, RS: Bookman.

How to Make a Directed Energy Weapon Detection System for Less Than $50. (n.d.). Instructables Circuits. Retrieved August 26, 2021, from https://www.instructables.com/id/How-to-Make-a-Directed-Energy-Weapon-Detection-Sys/?fbclid=IwAR3BwzQk3FqoiSFNPAVw-bRYGQQc2He8b09Pqb8yFiDLLdJbSzMNAUo3s2V4

Human auditory system response to modulated electromagnetic energy. (n.d.). Invasão e Controle Mental. https://invasaoecontrolemental.com.br/wp-content/uploads/2020/04/human-auditory-system-response-to-modulated-electromagnetic-energy.pdf

Human Remote Sensing - Human Spectral Imaging - Remote Biometrics. (n.d.). Human Remote Sensing - Remote Biometry. Retrieved August 25, 2021, from https://www.information-book.com/science-tech-general/human-remote-sensing-remote-biometry/

Interferência [Wave interference]. (n.d.). *In Wikiwand*. https://www.wikiwand.com/pt/Interfer%C3%AAncia

Invasão e Controle Mental. (2020, March 18). *Testes de reirradiação eletromagnéticas utilizando SDR*. [Video]. YouTube. https://www.youtube.com/watch?v=HLV5zegdHqU

Invasão e Controle Mental. (n.d.). *Home* [Invasão e Controle Mental]. YouTube. Retrieved August 24, 2021, from URL https://www.youtube.com/channel/UCQEwceYkANiF6PuBw1DPc7A

Ionosfera [Ionosphere]. (2020, September 19). *In Wikipedia*. https://pt.wikipedia.org/wiki/Ionosfera

Isaac Asimov. (n.d.). *In Wikiwand*. https://www.wikiwand.com/pt/Isaac_Asimov

Ivezic, M. (2018, April 30). *IEMI - Threat of Intentional Electromagnetic Interference*. 5G Security by Marin Ivezic. https://5g.security/cyber-kinetic/threat-of-iemi/?fbclid=IwAR3P9jeC-zoHxk8mPs_hazT_qOHNO1sz8xx2smwenaodEKExjK9y_vq0Hq-Q

Jacobs, J. (2019, March 26). *Trump Orders Study on Risks of Electromagnetic Weapon Attack*. Bloomberg. https://www.bloomberg.com/news/articles/2019-03-26/trump-is-said-to-plan-executive-order-on-electromagnetic-weapon?fbclid=IwAR1GNEMy1mseKflNVsMf-bAkl4PtZb3aoFwqTJKEuqA5By-XNwaujUMdyG9E

Jaspers, K. (2006). *Introdução ao pensamento filosófico* [Introduction to philosophical thought]. São Paulo, SP: Cultrix.

Jensen, B. (2012). *Microwave Instrument for Human Vital Signs Detection and Monitoring* [Doctoral dissertation, Technical University of Denmark]. https://backend.orbit.dtu.dk/ws/files/77581851/Brian_Sveistrup_Jensen_2012_Microwave_Instrument_for_Vital_Signs_Detection_and_Monitoring..PDF

Jiang, L., Stocco, A., Losey, D., Abernethy, J., Prat, C., & Rao, R. (2019, April 16). *BrainNet: A Multi-Person Brain-to-Brain Interface for Direct Collaboration Between Brains*. Scientific Reports. https://www.nature.com/articles/s41598-019-41895-7?fbclid=IwAR2Nhy5r4lMYRjshr-WiYDtL7tpXwCRuItdsUfEd8zNuzfpNF2kmUStV6EE8

José Manuel Rodríguez Delgado. (n.d.). *In Wikiwand*. https://www.wikiwand.com/en/Jos%C3%A9_Manuel_Rodriguez_Delgado

Joseph C. Sharp. (n.d.). *In Wikiwand*. https://www.wikiwand.com/en/Joseph_C._Sharp

Justesen, D. (1975). Microwaves and Behavior. *American Psychologist*.

Kato, R. (2017, November 14). Com novas técnicas, alemães conhecem melhor o cérebro [By using new techniques, Germans know the brain better]. *Exame*. https://exame.com/revista-exame/cerebro-a-fronteira-final/

Khan, A. (2017). *Microwave Engineering – Concepts and Fundamentals*. Boca Raton, Florida: CRC Press.

Laberge, S. (1985). *Sonhos Lúcidos* [Lucid Dreams]. Siciliano Livros, Jornais e Revistas Ltda.

Laika. (n.d.). *In Wikiwand*. https://www.wikiwand.com/pt/Laika

Laser. (2021, June 4). *In Wikipedia*. https://pt.wikipedia.org/wiki/Laser

Lent, R. (2015). *Neurociência da mente e do comportamento* [Neuroscience of mind and behavior]. Rio de Janeiro, RJ: Guanabara.

Li, X. P., Xia, Q., Qu, D., Wu, T., Yang, D., Hao, W., Jiang, X., & Li, X. M. (2014). *The Dynamic Dielectric at a Brain Functional...* Scientific Reports. https://www.nature.com/articles/srep06893?fbclid=IwAR2jmID1ELAPNz6GwFqgV336fcwyrSXnIHYadbT6GPKLSeAUzn-BFvwh5yQ&error=cookies_not_supported&code=ddee9446-31cb-4e4b-b01a-64a0843084ac

Lima, E. (2014). *Sistemas de Biometria de Frequência* [Frequency Biometric Systems]. Faculdade Integrada da Grande Fortaleza. Quixadá, CE.

Lima, L. (2020, January 27). *Os psicólogos que ensinaram a CIA "técnicas singulares" de tortura* [The psychologists who taught the C.I.A. how to torture]. BBC News. https://www.bbc.com/portuguese/internacional-51244029?at_medium=custom7&at_custom1=%5Bpost+type%5D&at_custom2=facebook_page&at_custom3=BBC+Brasil&at_campaign=64&at_custom4=3F606336-4103-11EA-A04C-BA0A3A982C1E&fbclid=IwAR2xtS1loW37CZMBrtJWN4l6fFiZKt02oXsUqOgSw-eQiwohTIBx4IpBO88

Lin, J. (1989). *Electromagnetic Interaction with Biological Systems*. New York: Plenum Press.

List of NRO launches. (2021, June 21). *In Wikipedia*. https://en.wikipedia.org/wiki/List_of_NRO_launches

List of spacecraft manufacturers. (2021, August 24). *In Wikipedia*. https://en.wikipedia.org/wiki/List_of_spacecraft_manufacturers

LSD. (n.d.). *In Wikiwand*. https://www.wikiwand.com/pt/LSD

Magnus Contact. (n.d.). *Mind Control – Mind Control, Neurotechnologies used as Weapons!* Mind Control. Retrieved August 25, 2021, from https://www.mindcontrol.se/

Marques, F. (2008). *Viabilidade de Implementação de um Sistema Biométrico de Autenticação* [Feasibility of Implementing a Biometric Authentication System]. Master's thesis, Universidade de Aveiro. Repositório Institucional da Universidade de Aveiro.

Martinovic, I., Davies, D, Frank, M., Perito, D., Ros, T., & Song, D. (n.d.). *On the Feasibility of Side-Channel Attacks with Brain-Computer Interfaces | USENIX*. USENIX Association. Retrieved August 24, 2021, from https://www.usenix.org/conference/usenixsecurity12/technical-sessions/presentation/martinovic

Martins, M., & Neves, I. (2015). *Propagação e radiação de ondas eletromagnéticas* [Propagation and radiation of electromagnetic waves]. São Paulo, SP: Lidel.

Matt Anderson. (2014, August 6). *EM Waves* [Video]. YouTube. https://www.youtube.com/watch?v=bwreHReBH2A

Mclean, C. (n.d.). *The Science of Microwaves Causing the Symptoms of the Diplomats in Cuba and China*. Google Docs. Retrieved August 24, 2021, from https://drive.google.com/file/d/1gyc6ETyzrrN5FxO47K2TX0cNGu_OFZbG/view?fbclid=IwAR2OANVyz-Mbk0PCQgqE0-hoNHajmzwdWvuzi7rKnCPN67Wki1FpmucDaTl

McPhate, M. (2016, June 10). *United States of Paranoia: They See Gangs of Stalkers*. The New York Times. https://www.nytimes.com/2016/06/11/health/gang-stalking-targeted-individuals.html?_r=0

Mecha. (2020, December 9). *In Wikipedia*. https://pt.wikipedia.org/wiki/Mecha

Mental Health Daily. (n.d.). *5 Types Of Brain Waves Frequencies: Gamma, Beta, Alpha, Theta, Delta*. Retrieved August 25, 2021, from https://mentalhealthdaily.com/2014/04/15/5-types-of-brain-waves-frequencies-gamma-beta-alpha-theta-delta/

Microwave auditory effect. (n.d.). *In Wikiwand*. https://www.wikiwand.com/en/Microwave_auditory_effect

Mlodinow, L. (2013). *Subliminar – Como o inconsciente influencia nossas vidas* [Subliminal: how the unconscious influences our lives]. Rio de Janeiro, RJ: Zahar.

Moreira, A. (1999, January). *Tecnologias de Transmissão* [Transmission Technologies]. DSI. http://www3.dsi.uminho.pt/adriano/Teaching/Comum/FactDegrad.html

MSNBC. (2018, September 11). *Russia Believed To Be Main Suspect In Attack On U.S. Diplomats | Velshi & Ruhle | MSNBC* [Video]. YouTube. https://www.youtube.com/watch?v=ghT-qxiI3yw&feature=youtu.be&fbclid=IwAR2IsOS7BRm_kxNoDhrO6TMgLRGErU3wOKo-Qly6TDraWiox2wk7Ay4Pw6Yo

Muxfeldt, P. (2017, April 26). *Propagação das ondas de rádio (802.11)* [Propagation of radio waves (802.11)]. CCM. https://br.ccm.net/contents/820-propagacao-das-ondas-de-radio-802-11

NA HORA DA GUERRA. (2017, May 19). *Guerra Eletrônica #6 - RADAR 2* [Video]. YouTube. https://www.youtube.com/watch?v=Y5uDnaRRYhg

NA HORA DA GUERRA. (2020, March 9). *Guerra Eletrônica #15 - RADAR Pulsado 1* [Video]. YouTube. https://www.youtube.com/watch?v=3YlE0YCBNSQ

NA HORA DA GUERRA. (n.d.). *Home* [NA HORA DA GUERRA]. YouTube. Retrieved August 24, 2021, from URL https://www.youtube.com/channel/UCkjw_BEpqjjy0O_mhfFbxDA

National Geographic Brasil. (2017, June). *Por que mentimos?* [Why do we lie?]. *National Geographic Brasil,* 207.

National Geographic Brasil. (2017, May). *Gênios* [Geniuses]. *National Geographic Brasil,* 206.

National Geographic Brasil. (2017, September). *O cérebro e os vícios* [The brain and addictions]. *National Geographic Brasil,* 210.

National Geographic Brasil. (2018, February). *Big brother da vida real* [Real life big brother]. *National Geographic Brasil,* 215.

National Reconnaissance Office. (n.d.). *In Wikiwand.* https://www.wikiwand.com/pt/National_Reconnaissance_Office

Nelson, B. (2019, April 25). *"Mind-Reading" Device Can Translate Your Brain Activity Into Audible Sentences*. Treehugger. https://www.treehugger.com/mind-reading-device-can-translate-your-brain-activity-audible-sentences-4862546

NetHoler. (2011, July 10). *Reincarnation - Airplane Boy (abc Primetime)* [Video]. YouTube. https://www.youtube.com/watch?time_continue=181&v=Uk7biSOzr1k&feature=emb_logo

Neurobiologia dos Sonhos: Atividade Elétrica [Neurobiology of Dreams: Electrical Activity]. (n.d.). Cérebro & Mente. Retrieved August 24, 2021, from http://www.cerebromente.org.br/n02/mente/neurobiologia.htm

Nicolelis, M. (2011). *Muito além do nosso eu* [Far beyond our ego]. Companhia Das Letras.

Nicolelis, M. (2012, April). *A monkey that controls a robot with its thoughts. No, really.* [Video]. TED Talks. https://www.youtube.com/watch?v=stXhGMVJuqA&list=PLus_DAOVXauYgyIgn-I6_7u2gUQEb-Rgu&index=33&t=0s

NKVD. (2021, August 20). *In Wikipedia.* https://en.wikipedia.org/wiki/NKVD

nptelhrd. (2013, September 16). *Lecture - 10 Single SideBand Modulation* [Video]. YouTube. https://www.youtube.com/watch?v=-ccrXpAJgjs&list=PLus_DAOVXauYgyIgn-I6_7u2gUQEb-Rgu&index=28&t=434s

Oliveira, A. (2019, August 9). *Este dispositivo ouve a voz que fala dentro da sua cabeça* [This device listens to the voice that speaks inside your head]. *Super Interessante.* https://super.abril.com.br/ciencia/este-dispositivo-ouve-a-voz-que-fala-dentro-da-sua-cabeca/?fbclid=IwAR31ME02gSzhj7h5GGXGUZ0KhIVu2EKPw-p1LMtK4p8wf2GKHgTyg88Aexw

Oliveira, R. (2000). *Neurolingüística e o aprendizado da linguagem* [Neurolinguistics and language learning]. Brasília, DF: Respel.

Open Source Tools for Neuroscience. (n.d.). OpenBCI. Retrieved August 25, 2021, from https://openbci.com/?fbclid=IwAR0a93kXoMC4iPNvx3G0sd43Rzt0yRpD90YSQo2pE-FpNM6sqO_CK-YJqGbU

Os Tipos de Memória – Memorização [The Types of Memory – Memorization]. (n.d.). Memorização. Retrieved August 25, 2021, from https://memorizacao.info/os-tipos-de-memoria.html

Pall, M. (2015). *Microwave frequency electromagnetic fields (EMFs) produce widespread neuropsychiatric effects including depression.* Journal of Chemical Neuroanatomy. Retrieved August 25, 2021, from URL https://www.researchgate.net/publication/281261829_Microwave_frequency_electromagnetic_fields_EMFs_produce_widespread_neuropsychiatric_effects_including_depression?fbclid=IwAR3SYoNPJDOjuFPg906XtihdwTDoKw-lR-PGuuPsm3eLbP7tYLMpcyDb2BY

Pedofilia [Pedophilia]. (2021, August 23). *In Wikipedia.* https://pt.wikipedia.org/wiki/Pedofilia

Peer-to-peer. (2021, August 11). *In Wikipedia.* https://pt.wikipedia.org/wiki/Peer-to-peer

Pelley, S. (2019, September 1). *Brain trauma suffered by U.S. diplomats in Cuba, China could be work of hostile foreign government.* CBS News. https://www.cbsnews.com/news/brain-trauma-suffered-by-u-s-diplomats-abroad-could-be-work-of-hostile-foreign-government-60-minutes-2019-09-01/?fbclid=IwAR3LdAEr97701pU0VhvPpVJYSSYwnwFN2gPxKwG-F28ckLgJH-SoaliMIsMY

Pennicott, K. (2002, May 16). *Noisy signals strengthen human brainwaves.* Physics World. https://physicsworld.com/a/noisy-signals-strengthen-human-brainwaves/?fbclid=IwAR0Of6v5k-nK2xnsAOkOft99DsasXbRBeLnjsxmMq3-9UByuk2e3DBl7Aw

Penttinen, J. (2015). *The Telecommunications Handbook - Engineering Guidelines for Fixed, Mobile and Satellite Systems.* Hoboken, NJ: Wiley.

Perper, R. (2019, July 17). *Elon Musk's company Neuralink plans to connect people's brains to the internet by next year using a procedure he claims will be as safe and easy as LASIK eye surgery*. Insider. https://www.businessinsider.com/elon-musk-neuralink-implants-link-brains-to-internet-next-year-2019-7?fbclid=IwAR1br4lcC8DswkwtYt2JiMBTShmMJ3kzl-trdx7yLYAbjiPPXzXWmMMAfmzI

Persinger, M. (1974). *ELF and VLF Electromagnetic Field Effects*. New York: Plenum Press.

Pilati, R. (2018). *Ciência e Pseudociência – Por que acreditamos naquilo em que queremos acreditar* [Science and Pseudoscience – Why we believe what we want to believe]. São Paulo, SP: Editora Contexto.

Pisadeira (folclore) [*Pisadeira* (folklore)]. (2021, July 1). In *Wikipedia*. https://pt.wikipedia.org/wiki/Pisadeira_(folclore)

Portilho, G. (2016, December 15). *Como funciona o olho humano?* [How does the human eye work?]. *Super Interessante*. https://mundoestranho.abril.com.br/saude/como-funciona-o-olho-humano/

Primatas [Primate]. (2021, May 8). *In Wikipedia*. https://pt.wikipedia.org/wiki/Primatas

Processos de Moscou [Moscow Trials]. (2021, August 13). *In Wikipedia*. https://pt.wikipedia.org/wiki/Processos_de_Moscou

Projeto Escrita Criativa. (n.d.). *Home* [Projeto Escrita Criativa]. YouTube. Retrieved August 25, 2021, from URL https://www.youtube.com/channel/UCO6nYvm-muWS5rR1rpz00eA

Qual a diferença entre uma transmissão AM para uma FM? [What is the difference between AM and FM?] (2012, October 26). A Física do Rádio. http://fisica3ufrb.blogspot.com/2012/10/qual-diferenca-entre-uma-transmissao-am.html

Radar. (n.d.). *In Wikiwand*. https://www.wikiwand.com/pt/Radar

RADIO WAVES below 22 kHz. (n.d.). RADIO WAVES below 22 KHz. Retrieved August 26, 2021, from http://www.vlf.it/

Radioescuta DX. (n.d.). *Bem-vindo a Radioescuta e DX* [Welcome to Radioescuta and DX]. Retrieved August 25, 2021, from http://www.sarmento.eng.br/OndasCurtas.htm

Redação Galileu. (2019, January 11). *"Som misterioso" ouvido por diplomatas em Cuba era feito por grilos* [A 'mysterious sound' heard by diplomats at the U.S. embassy in Cuba was caused by crickets]. Galileu. https://revistagalileu.globo.com/Sociedade/noticia/2019/01/som-misterioso-ouvido-por-diplomatas-em-cuba-era-feito-por-grilos.html

Rede neural artificial [Artificial neural network]. (n.d.). *In Wikiwand.* https://www.wikiwand.com/pt/Rede_neural_artificial

Relé [Relay]. (2020, April 19). *In Wikipedia.* https://pt.wikipedia.org/wiki/Rel%C3%A9

Resnick, B. (2016, June 20). *Scientists have invented a mind-reading machine. It doesn't work all that well.* Vox. https://www.vox.com/2016/6/20/11905500/scientists-invent-mind-reading-machine

Resolução comentada dos exercícios de vestibulares sobre Polarização e Ressonância de ondas [Annotated answers of college entrance exam exercises on Polarization and Wave Resonance]. (n.d.). Física e Vestibular - Aulas Grátis de Física. http://fisicaevestibular.com.br/novo/ondulatoria/ondas/polarizacao-e-ressonancia-de-ondas/resolucao-comentada-dos-exercicios-de-vestibulares-sobre-polarizacao-e-ressonancia-de-ondas/

Riera, A., Dunne, S., Cester, I., & Ruffini, G. (2008). *STARFAST: a wireless wearable EEG/ECG biometric system based on the ENOBIO Sensor.* https://www.yumpu.com/en/document/read/53643357/starfast-a-wireless-wearable-eeg-ecg-biometric-system-based-on-the-enobio-sensor

Riera, A., Soria-Frisch, A., Caparrini, M., Grau, C., & Ruffini, G. (2007). Unobtrusive Biometric System Based on Electroencephalogram Analysis. *EURASIP Journal on Advances in Signal Processing.* https://doi.org/10.1155/2008/143728

Rocha, A. (2009*). Análise das respostas eletrofisiológicas de longa latência – P300 em escolares com e sem sintomas de Transtorno do Processamento Auditivo* [Long latency evoked responses analysis in school-aged children with and without auditory processing disorders symptoms]. Final paper, Universidade Federal de Minas Gerais. FTP Medicina – UFMG. https://ftp.medicina.ufmg.br/fono/monografias/2009/anapaula_analisedasrespostas_2009-1.pdf

Ross Adey (1922–2004). (2004, May 20). Microwave News. https://microwave-news.com/news-center/ross-adey

Ross, B., Schwartz, R., Meek, J., & Kreider, R. (2017, January 7). *What We Know About Esteban Santiago, Suspect in Fort Lauderdale Attack.* ABC News. https://abcnews.go.com/US/esteban-santiago-suspect-fort-lauderdale-attack/story?id=44612498

Ross, C. (n.d.). *Project Bluebird.* Want to Know. Retrieved August 24, 2021, from https://www.wanttoknow.info/bluebird10pg

Ruminantes [Ruminant]. (2021, January 7). *In Wikipedia.* https://pt.wikipedia.org/wiki/Ruminantes

Saffi, F., & Serafim, A. (2015). *Neuropsicologia forense* [Forensic neuropsychology]. Porto Alegre, RS: Artmed.

Saleem, A. (2021, August). *SDR for Ethical Hackers and Security Researchers* [Udemy course]. Udemy. https://www.udemy.com/share/101tjIBU-cZdldRR34=/

Sanei, S., & Chambers, J. (2007). *EEG Signal Processing.* Hoboken, New Jersey: Wiley.

Santos, K. (2011). *Magnetron: do radar ao forno de micro-ondas* [Magnetron: from radar to the microwave oven]. Universidade Católica de Brasília. https://livrozilla.com/doc/282438/magnetron--do-radar-ao-forno-de-micro-ondas

Satélite artificial [Satellite]. (2021, August 4). In *Wikipedia*. https://pt.wikipedia.org/wiki/Sat%C3%A9lite_artificial

Satélite artificial [Satellite]. (n.d.). In *Wikiwand*. https://www.wikiwand.com/pt/Sat%C3%A9lite_artificial

Scherer, A. (2018, February 21). *A química da mente produtiva* [The chemistry of the productive mind]. *Exame*, 1151.

Schmidt, M. (2013, September 25). *Gunman Said Electronic Brain Attacks Drove Him to Violence, F.B.I. Says*. The New York Times. https://www.nytimes.com/2013/09/26/us/shooter-believed-mind-was-under-attack-official-says.html

Scientific American Brasil. (2018, July). *Prevenção ao suicídio* [Suicide prevention], *Scientific American Brasil*, 184, 59.

Scientific American Brasil. (2019a, March). *O Código Facial* [The Facial Code]. *Scientific American Brasil*, 193.

Scientific American Brasil. (2019b, March). *Robôs que aprendem sozinhos* [Robots that learn on their own]. *Scientific American Brasil*, 181.

SDR = Software-Defined Radio. (2016, May 16). SDRZero. https://www.qsl.net/py4zbz/sdr/sdrz.htm

Sebrae. (2017, April 11). *Definição de Patente* [Patent Definition]. https://www.sebrae.com.br/sites/PortalSebrae/artigos/definicao-de-patente,230a634e2ca62410VgnVCM100000b272010aRCRD

Sharp, J. (1973). Voice to Skull Demonstration: Artificial microwave voice to skull transmission was successfully demonstrated by researcher Dr. Joseph. *American Psychologist*.

Sharp, J. (1974). *Artificial microwave voice to skull transmission was successfully demonstrated*. Seminar from the University of Utah.

Silva, M. (n.d.). *Nervo óptico* [Optic nerve]. InfoEscola. Retrieved August 25, 2021, from https://www.infoescola.com/visao/nervo-optico/

Silva, V., Pereira, J., Nohara, E., & Rezende, M. (n.d.). *Comportamento eletromagnético de materiais absorvedores de micro-ondas baseados em hexaferrita de Ca modificada com íons CoTi e dopada com La* [Electromagnetic behavior of radar absorbing materials based on Ca hexaferrite modified with Co-Ti ions and doped with La]. SciELO. Retrieved August 25, 2021, from http://www.scielo.br/scielo.php?pid=S2175-91462009000200255&script=sci_abstract&tlng=pt

Silve, S. (1984). *Microwave Antenna Theory and Design (Electromagnetics and Radar)*. New York, NY: Institute of Electrical Engineers.

simonxhayes. (2009, July 26). *Computer records animal vision in Laboratory – UC Berkeley* [Video]. YouTube. https://www.youtube.com/watch?v=piyY-UtyDZw

Snyder, B. (2019, November 1). *The Next Computer Revolution Will Be Based on Our Brains*. Bloomberg. https://www.bloomberg.com/news/articles/2019-11-01/how-the-human-brain-project-aims-to-improve-the-world-s-computers?fbclid=IwAR06wom4CeHJnnnqZk7QqX-0RuMPxSDkRfVJI-l40nX8xeaMMDHa0_sjigA

Soro da verdade [Truth serum]. (2019, May 23). *In Wikipedia*. https://pt.wikipedia.org/wiki/Soro_da_verdade

Stephen Hawking. (n.d.). *In Wikiwand*. https://www.wikiwand.com/pt/Stephen_Hawking

STOPEG - Stop Electronic Weapons and Gang Stalking. (n.d.). STOPEG Foundation. Retrieved August 25, 2021, from https://www.stopeg.com/

Study: Women Need More Sleep Because Of One Obvious Reason. (n.d.). SimpleCapacity. Retrieved August 25, 2021, from https://simplecapacity.com/2015/11/study-women-need-more-sleep-because-of-one-obvious-reason/

Suçuarana, M. (n.d.). *Formigas zumbis* ['Zombie ants']. InfoEscola. Retrieved August 25, 2021, from https://www.infoescola.com/biologia/formigas-zumbis/

Super Interessante. (2016, October). *Mindfulness: como domar a sua mente agora* [Mindfulness: how to tame your mind]. *Super Interessante*, 365.

Super Interessante. (2017, June). *Cérebro* [Brain]. *Super Interessante*, 357.

Super Interessante. (2019, May). *A Ciência das Emoções* [The Science of Emotions]. *Super Interessante*, 396.

Superposição de ondas (continuação) [Wave superposition]. (n.d.). Só Física. Retrieved August 24, 2021, from https://www.sofisica.com.br/conteudos/Ondulatoria/Ondas/superposicao2.php

Suppes, P., Lu, Z.-L., & Han, B. (1997). Brain wave recognition of words. *PNAS*. https://www.pnas.org/content/pnas/94/26/14965.full.pdf

TecMundo. (2016, February 11). *HAARP: o projeto militar dos EUA que pode ser uma arma geofísica* [HAARP: The U.S. military project that could be a geophysical weapon]. https://www.tecmundo.com.br/tecnologia-militar/8018-haarp-o-projeto-militar-dos-eua-que-pode-ser-uma-arma-geofisica.htm

TED. (2013, September 23). *Elizabeth Loftus: A ficção da memória* [Video]. YouTube. https://youtu.be/PB2OegI6wvI

TED. (2019a, June 3). *A new way to monitor vital signs (that can see through walls) | Dina Katabi* [Video]. YouTube. https://www.youtube.com/watch?v=CXy1by-guvJY&list

TED. (2019b, June 3). *Sleep is your superpower | Matt Walker* [Video]. YouTube. https://www.youtube.com/watch?v=5MuIMqhT8DM

TED. (2020, June 14). *Can we edit memories? | Amy Milton* [Video]. YouTube. https://www.youtube.com/watch?v=ZK7ih4V0erc

TEDx Talks. (2017, August 29). *New Brain Computer interface technology | Steve Hoffman | TEDxCEIBS* [Video]. YouTube. https://www.youtube.com/watch?v=CgFzmE2fGXA

Teleco. (n.d.). *Dados na Rede Celular: Evolução das Tecnologias* [Data on the Cellular Network: Development of Technologies]. Retrieved August 25, 2021, from https://www.teleco.com.br/tutoriais/tutorialtrafdados/pagina_2.asp

Texas Instruments. (2018, January 8). *People counting demonstration using TI mmWave sensors* [Video]. YouTube. https://www.youtube.com/watch?v=RT56YzqME6M&list=PLus_DAOVXauYgyIgn-I6_7u2gUQEb-Rgu&index=48&t=7s

Texas Instruments. (2019, August 20). *Intelligent Fall Detection Using TI mmWave Sensors* [Video]. YouTube. https://www.youtube.com/watch?v=njhR-wijx_HY&list=PLus_DAOVXauYgyIgn-I6_7u2gUQEb-Rgu&index=47&t=0s

The Journal of Nervous and Mental Disease. (1968). *ICD-10 classification of mental and behavioral disorders: clinical descriptions and diagnostic guidelines.* World Health Organization. Geneva: World Health Organization.

The New York Times. (1979, May 30). *Soviet Halts Microwaves Aimed at U.S. Embassy.* https://www.nytimes.com/1979/05/30/archives/soviet-halts-microwaves-aimed-at-us-embassy.html

The New York Times. (2019, October 15). *Are We Ready for Satellites That See Our Every Move?* https://www.nytimes.com/2019/10/15/opinion/satellite-image-surveillance-that-could-see-you-and-your-coffee-mug.html?fbclid=IwAR3ISFsElpTiBA-iDBvXDg0tch7GR0349qkiLlhwWF9ml--6_a5oCu2VeZo

The Science of Electronic Harassment. (2020, August 15). *Detection of frequency ranges used for electronic harassment.* [Video]. YouTube. https://www.youtube.com/watch?v=sP4ZjyrfuUo&

The Science of Electronic Harassment. (n.d.). Home [The Science of Electronic Harassment]. YouTube. https://www.youtube.com/channel/UCl4nSNVf7uJPsekm-nUWpBmw

Thorpe, J., Oorschot, P., & Somayaji, A. (2005). Pass-thoughts: authenticating with our minds. *Proceedings of the 2005 Workshop on New Security Paradigms - NSPW '05*, 45–56. https://doi.org/10.1145/1146269

Toscano, R. (2006). *Bloqueador de múltiplas frequências: concepção do sistema e estudo de caso para terminais is-95* [Multiple-frequency blocking: system design and case study for is-95 terminals]. Master's thesis, Instituto Militar de Engenharia. Programa de Pós-Graduação em Engenharia Elétrica.

Transponder. (n.d.). *In Wikiwand.* https://www.wikiwand.com/pt/Transponder

Transtorno de Personalidade Histriônica ou Histérica (TPH) [Histrionic personality disorder]. (2010, August 17). Memórias de uma Methamorfose. http://memoriasdeumamethamorfose.blogspot.com/2010/08/transtorno-de-personalidade-histrionica.html

UNICAMP. (n.d.). *Córtex motor normal em hematoxilina - eosina (HE). Neurônios piramidais, células gigantes de Betz* [Normal motor cortex with hematoxylin-eosin (HE). Pyramidal neurons, giant Betz cells]. Anatpat – UNICAMP. Retrieved August 25, 2021, from http://anatpat.unicamp.br/bineu-cortexmotornlhe.html

Universo Programado. (2020, June 25). *Inteligência Artificial detecta humanos ATRAVÉS de paredes!* [Video]. YouTube. https://www.youtube.com/watch?v=JWuS6q9EYAo&list=PLus_DAOVXauYgyIgn-I6_7u2gUQEb-Rgu&index=15&t=0s

Universo Programado. (n.d.). *Home* [Universo Programado]. YouTube. Retrieved August 25, 2021, from URL https://www.youtube.com/channel/UCf_kacKyoR-RUP0nM3obzFbg

UNIVESP. (2017, December 6). *Eletromagnetismo – Apresentação da disciplina* [Video]. YouTube. https://www.youtube.com/watch?v=-UQGaneAZW8

UNIVESP. (2018, August 7). *Licenciatura em Física - Eletromagnetismo - 14º Bimestre* [Video]. YouTube. https://www.youtube.com/playlist?list=PLxI8Can9yAHfsS-KveLkqvvO3yZrGrNiQO

Veja. (2016, May 6). *Obama lança programa para mapear cérebro humano* [Obama launches program to map the human brain]. *Veja*. https://veja.abril.com.br/ciencia/obama-lanca-programa-para-mapear-cerebro-humano/

Velho, J. (2016). *Tratado de Computação Forense* [Computer Forensic Treaty]. Campinas, SP: Millennium.

Vicente, J. (2019, November 1). *Controlar máquinas com o pensamento parece um sonho, mas pode ser pesadelo* [Controlling machines with thought sounds like a dream, but it can be a nightmare]. Tilt UOL. https://www.uol.com.br/tilt/noticias/redacao/2019/11/01/controlar-maquinas-com-o-pensamento-parece-um-sonho-mas-pode-ser-pesadelo.htm?fbclid=IwAR1p2xNuO-NgL3gYr-NYymDhWWWslt0d8s_XkBleaH2D5JWNcnurBD1TzzI0

Vigotskii, L., Luria, A., & Leontiev, A. (2001). *Linguagem, desenvolvimento e aprendizagem* [Language, development and learning]. São Paulo, SP: ícone.

Vigotsky, L. (2010). *A construção do pensamento e da linguagem* [The construction of thought and language]. São Paulo, SP: Martins Fontes.

Vilicic, F., & Thomas, J. (2021, March 26). *Os trunfos e os riscos da inteligência artificial* [Benefits and risks of artificial intelligence]. *Veja*. https://veja.abril.com.br/tecnologia/os-trunfos-e-os-riscos-da-inteligencia-artificial/

Voltolini, R. (2014, December 11). *China cria arma de micro-ondas que "ferve" moléculas de água do corpo* [China creates microwave weapon that "heats" water molecules in the body]. TecMundo. https://www.tecmundo.com.br/armas-de-fogo/69208-china-cria-arma-micro-ondas-ferve-moleculas-agua-corpo.htm?fbclid=IwAR1IJHcP1xY0LloEW3K-Fu-Heg1jgz4yhkIfmZWxYJzkWBTbdWwiyGP6jXeI

Wang, J.-K., Jiang, X., Peng, L., Li, X.-M., An, H.-J., & Wen, B.-J. (2019, January). *Detection of Neural Activity of Brain Functional Site Based on Microwave Scattering Principle.* Department of Information and Communication, Guilin University of Electronic Technology. Retrieved August 25, 2021, from URL https://www.researchgate.net/publication/330548074_Detection_of_Neural_Activity_of_Brain_Functional_Site_Based_on_Microwave_Scattering_Principle?fbclid=IwAR3pCaVVyuLrq-2LUry09dTFDgPqlSOyi9JOK-pBe5TD4_DcCmDbtPx9YpnA

WannaCry [WannaCry ransomware attack]. (2020, April 27). In Wikipedia. https://pt.wikipedia.org/wiki/WannaCry

White, D. (2019, March 25). *'Google brain' implants could mean end of school as anyone will be able to learn anything instantly.* The Sun. https://www.thesun.co.uk/tech/8710836/google-brain-implants-could-mean-end-of-school-as-anyone-will-be-able-to-learn-anything-instantly/

Wolpert, D. (2011, July). *The real reason for brains*. TED Talks. https://www.ted.com/talks/daniel_wolpert_the_real_reason_for_brains#t-77064

World Science Festival. (2015, March 18). *The Mind After Midnight: Where Do You Go When You Go to Sleep?* [Video]. YouTube. https://www.youtube.com/watch?v=stXhG-MVJuqA&list=PLus_DAOVXauYgyIgn-I6_7u2gUQEb-Rgu&index=33&t=0s

The more than 500 references used to support the research for the preparation of this book will soon be available on the official website:

www.invasionandmindcontrol.com